The Water Sensitive City

The Water Sensitive City

Gary Grant

WILEY Blackwell

Library of Congress Cataloging-in-Publication data applied for

ISBN: 9781118897669

A catalogue record for this book is available from the British Library.

Wiley also publishes its books in a variety of electronic formats. Some content that
appears in print may not be available in electronic books.

Set in 10/12.5pt Avenir by SPi Global, Pondicherry, India
Printed and bound in Singapore by Markono Print Media Pte Ltd

1   2016

Contents

About the Author

Gary Grant is a Chartered Environmentalist, Fellow of the Chartered Institute of Ecology and Environmental Management, thesis tutor at the Bartlett Faculty of the Built Environment, University College London and Director of the Green Infrastructure Consultancy (formerly the Green Roof Consultancy). In 2006 he wrote *Green Roofs and Facades* published by BRE Press and in 2012, *Ecosystem Services Come to Town – Greening Cities by Working with Nature*, published by Wiley-Blackwell. From 2006 to 2009 he was a director of EDAW and then AECOM Design + Planning, where he worked on large-scale planning projects including the London 2012 Olympic Park, the Bedford Valley River Park, the Whitehill-Bordon eco-town, Education City, Qatar and Saadiyat Island, Abu Dhabi. More recently, with the Green Infrastructure Consultancy, he has been working on planning and design, including green infrastructure networks for cities, green roofs, living walls and rain gardens.

Acknowledgement

I would like to thank my wife Sue for her support and understanding during the writing of this book.

1. Water and Cities

The Molecule

Water is remarkable. It is an odourless, tasteless and transparent molecule. Consisting of two hydrogen atoms bonded to a single oxygen atom, with each water molecule weakly connected to its neighbour, water is a relatively sticky liquid, with a high boiling point compared to other species of molecule of a similar atomic mass. Liquid water forms a solvent, solute and reactant that channels life. As far as we know, biological reactions do not occur in the absence of water. Barring new supplies delivered in the form of comets (an extremely infrequent occurrence fortunately), the amount of water on earth remains constant.[1]

Blue Planet

We inhabit a watery blue planet. When viewed from space, the oceans give our only home its blue colour. Earth is predominantly blue, but also white – with the white caps of the polar ice and the swirling white clouds organized into weather systems. Water, whether seen by astronauts, or viewed by the earthbound, may appear to be abundant, however it constitutes, in effect, a thin film on the surface of the planet. If the water of the earth, all 1.386 million cubic kilometres of it, were to be put into a single drop, it would create a sphere only 1384 km in diameter. To put this in context, the diameter of the earth is 12,742 km.

The Water Sensitive City, First Edition. Gary Grant.
© 2016 John Wiley & Sons, Ltd. Published 2016 by John Wiley & Sons, Ltd.

For a sense of scale, compare a marble (equivalent to the volume of all the water of the earth) with a basketball (equivalent to the volume of the earth). The saltwater of the oceans makes up 96.5% of the total reservoir of water, the rest being groundwater, vapour, rivers, lakes and ice. Most freshwater, about 24 million cubic kilometres of it, is locked up in glaciers and ice caps. 10.5 million cubic kilometres of freshwater occurs as groundwater, with less than 200,000 km^3 of water in lakes, rivers and wetlands. Readily available liquid freshwater in rivers and lakes totals 93,113 km^3 and could be contained in a sphere just 56.2 km in diameter.[2] Only about 2.5% of the earth's water is suitable for human consumption without some kind of treatment. Water is ubiquitous in the biosphere; yet clean, safe, drinkable, freshwater is a relatively scarce resource.

A Global Water Cycle

Water moves and changes state as part of a perpetual planetary hydrological cycle. Radiation from the sun, striking the earth as it revolves, heats seas, lakes, soil and vegetation, causing water to evaporate. The sun also drives plant transpiration, the process whereby water passes through plants and exits via the leaves. As night turns to day and parts of the earth turn to face the sun, the warming water vapour forms into clouds. These clouds then move through the atmosphere in a process known as advection. When the temperature of the air falls, as it meets colder air, or as it cools when it rises, the water in clouds condenses and falls as rain, sleet or snow. As day turns to night and the dark side of the earth cools, dew may form (often the only source of water for the denizens of the desert). Where snow falls onto ice caps and glaciers it may accumulate and be sequestrated for millennia. Spring melt, by contrast may come from snow that has lain for no more than a few days, weeks or months. Rain falls back to the oceans or onto the land. It may be intercepted by vegetation, never reaching the ground, or may infiltrate into the soil. Surplus rainfall forms surface or underground flows, entering lakes, streams and rivers, with the latter usually reaching the oceans. Where soil is saturated or frozen, or where soil or rocks are impermeable, rainfall will form runoff and enter water courses. In locations where the geological conditions are suitable, where the rocks are permeable, water replenishes aquifers, where in some cases, like the water of the ice caps, it may remain for millennia – the so-called fossil waters.[3]

Terrain and Water

Topography, geology and biomes[4] have strong influences over where water collects and flows. High ground stimulates clouds to produce rainfall as the clouds are pushed upwards into colder air by prevailing winds. The leeward sides of mountains may receive less rainfall and are therefore said to fall within rain shadows. The land divides along watersheds into river basins or catchments, where rain and snow melt feed particular river systems, forests and wetlands. Small catchments have small rivers and cannot support large settlements by themselves. Large rivers, like the Nile, Indus, Tigris, Euphrates and Yellow River, carry silt that was the foundation of agricultural systems that supported the first cities and civilizations. Humans continue to modify the water cycle and those modifications have been increasing in extent and intensity, particularly since the middle of the twentieth century. There are particular problems with those places where people are exploiting the upper parts of catchments, intercepting or diverting freshwater that would otherwise supply communities downstream, a problem that is predicted to lead to an increase in conflict and even warfare between nations.[5] In addition, poor management practices, for example, deforestation in the upper reaches of river basins or an overreliance on piped drainage, can also lead to flooding and pollution problems downstream. Integrated catchment (river basin) management is frequently and quite rightly promoted as best practice but is usually applied in an inadequate and unsatisfactory way because of administrative and political divisions, conflicting private and public interests

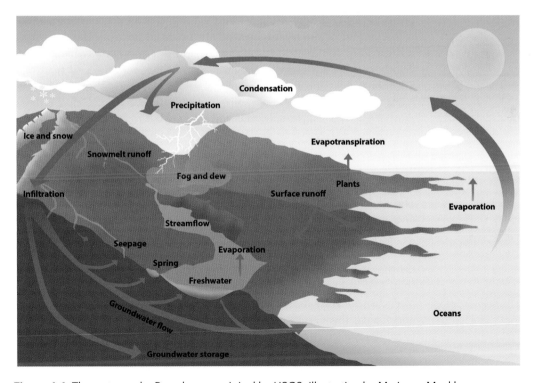

Figure 1.1 The water cycle. Based on an original by USGS. Illustration by Marianna Magklara.

or just plain ignorance. Watersheds (also known as river basins or catchments) would make the ideal administrative boundaries, but catchments frequently traverse administrative, political and even national boundaries, making comprehensive integrated catchment management plans difficult to agree and implement.

Seasons and Cycles

The 23.5° tilt of the earth's axis results in the northern hemisphere being more exposed to the sun from May to July and the southern hemisphere being more exposed to the sun from November to January. These annual changes bring the colder and wetter weather of winter to temperate regions and the wet (monsoon) seasons in the tropics. There is a larger landmass and therefore more plant biomass in the northern hemisphere, which means that the global atmospheric carbon dioxide concentration fluctuates, falling during the northern summer as plants grow and absorb carbon dioxide and increasing again through the northern winter as plant growth slows and, in some cases, halts. The current overall trend of atmospheric carbon dioxide concentration, of course, is up – largely the result of the burning of fossil fuels. The oceans play a key role in modifying the climate because they absorb and store heat. Ocean temperatures affect atmospheric temperatures, oceans currents and wind and the Pacific Ocean, which is the largest ocean by far, has the strongest impact of global weather patterns, as demonstrated by the El Nino phenomenon, which causes floods and drought across the Americas and as far afield as Australia, Southeast Asia and Africa.[6] Seasonal effects mean that rainfall in most parts of the world is uneven, with many regions experiencing intense rainfall for short periods followed by extended dry spells.

Variations in Rainfall

The amount of rain that falls varies considerably from region to region and place to place. For example, the heaviest rains of more than 11,000 mm per year occur where monsoon clouds meet the Kharsi Hills on the slopes of the eastern Himalayas in north-east India. Vancouver, on the rainy northwest Pacific coast of North America, enjoys more than 1100 mm of rainfall per year. London, England, to the surprise of many, is relatively dry, receiving only 600 mm of precipitation per year and Cairo, the capital of Egypt, receives just 25 mm of rainfall each year.[7] Rainfall patterns can be unpredictable. Even places noted for their reliable rainy season, like Ecuador for example, can suffer drought. In 2009, during an El Nino event, that country suffered its worst drought for 40 years.[8] As a result of the drought, reservoirs dried up, leading to water shortages in the cities, however much of the news at the time was dominated by stories of power blackouts, caused because of the lack of water to drive the turbines of the country's hydroelectric power stations.[9]

Changing Climates

As climate changes, so does the water cycle; 25,000 years ago, during the last ice age, sea levels were 120 m lower than at present, with more water locked up in the polar ice caps and mountain glaciers. The Ice Age climate of that time was drier and rainfall was lower overall than it is at present. Rainforests shrank in size and deserts and grasslands expanded.[10] As global temperatures warmed after the end of the last Ice Age, the atmosphere increased its capacity to hold water vapour, in turn changing weather patterns, which then allowed both tropical and temperate forests to expand in area. Anthropogenic (man-made) climate change is accelerating the process of warming, with the ice caps and mountain glaciers shrinking still further and sea levels rising. The atmosphere is predicted to carry even more water, bringing more unsettled weather with heavier downpours, more powerful storms and longer droughts. (Read more on climate and climate change in Chapter 5.)

Atmospheric Carbon Dioxide

There has been increase in atmospheric carbon dioxide caused by deforestation, agricultural intensification and expansion and, more recently, the burning of fossil fuels (an increase from 280 parts per million in the year 1800 to 400 parts per million in 2015).[11] This has had indirect effects on the water cycle but there have also been direct impacts. Deforestation, which usually leads to the creation of new pastures or croplands, tends to dry out soils and the landscape as a whole. Following deforestation, there are increases in surface runoff and therefore overall reductions in the volume of water evaporated and reductions in quantities of ground water. Regional patterns of cloud formation, and therefore rainfall, also change. Once denuded of forest vegetation, soils lose some of their organic matter and associated capacity to store water. The problem is further exacerbated as wetlands are also drained to create farmland. Then the farmland itself is drained. When this occurs, organic matter is oxidized and carbon dioxide is released into the atmosphere. Where crops, which require large quantities of water, are introduced, irrigation often becomes necessary, resulting in the unsustainable exploitation of groundwater or overabstraction of water from rivers. Globally, around 70% of the water abstracted from rivers, wells and boreholes is used for agriculture.[12] Lake-fed rivers (like, for example, the Aral Sea) shrink or may disappear altogether as the result of abstraction of water for agricultural use.[13] Excessive irrigation in arid climates may also result in increased soil salinity, which can inhibit plant growth and lead to a significant reduction the range of crop species that may be grown. In some cases land may be abandoned as the result of salinification.[14]

Fossil Fuels and Growth

Fossil fuels powered the Industrial Revolution. The world's population grew steadily from a billion in 1800 to 2 billion in 1920 – unprecedented growth, in effect powered by coal – however, even more dramatic change came with the onset of the Oil Age, with an increase in population from 2 billion to 7 billion people during the 90 years between 1920 and 2010. The global population is still growing and is predicted to peak at around 9 or 10 billion by 2050, a further increase of 2 to 3 billion. Global population growth has also been a story of urbanization and mechanization. The Industrial Revolution reduced the demand for farm labour as agriculture became increasingly mechanized. There was also a demand for labour to man the new factories, a demand that also drove the migration of people from countryside to town. This, in turn, caused towns and cities to grow rapidly – a process that still continues in developing countries. The population of Manchester, an industrialized city in the northwest of England, for example, grew from around 330,000 in 1800 to more than 2.5 million people in 1920. The population of Rio de Janeiro in Brazil increased from about 500,000 in 1900 to its current level of more than 6 million, with similar numbers of people in the immediate hinterland. These increases in city populations have been repeated and are still being repeated all over the world, so that now more than 50% of the world's population lives in urban areas. In developed countries the vast majority of the population is already urban. This trend looks set to continue, perhaps until after the global population peaks later this century. Across the world, on average, 5 million people move to cities every month. Water demand thereby increases – water for the agriculture that feeds the populations of the cities and water to supply the people in their dwellings and places of work. Increases in incomes change lifestyles, with more bathing and an increase in ownership of water-consuming equipment and processes. (See Chapter 3 for more information on why the demand for freshwater is increasing.)

The Ancients and Water

The first city dwellers relied on springs or wells for most of their supplies of potable water, but would often supplement this with rainwater collected from roofs and subsequently directed into purpose-built cisterns (storage tanks). For example, large cisterns holding $50\,m^3$ or more, dating back to the second millennium BC, have been described from Minoan sites.[15] Per capita water use was low during this period and sizeable communities of tens of thousands could be supported in this way; however, as cities grew still further, water needed to be brought from further afield. The ancient Romans, for example, who numbered in total approximately 1 million people, constructed a series of aqueducts to bring water from distant upland springs and streams.[16] This trend, of bringing water to cities from increasingly distant upland locations, has continued and accelerated to the present day. The combination of more urban dwellers, each consuming more water, means that growing volumes of high-quality freshwater water need to be directed to cities, drying out the upper sections of river catchments and polluting the lower sections of rivers and coastal seas.

Dams

Rivers are dammed, sometimes many times, to create reservoirs for irrigation and drinking water but also for the generation of electricity using hydro-electric power stations. Dams are nearly always constructed before the affected aquatic ecosystems are properly described and the full spectrum of ecological impacts is fully understood. Dams block the migration of fish, including some species like salmon, which spawn in the shallow fast-flowing streams that are often occur upstream of dams. Another major consequence of dam building is that sediments are trapped behind the dam, rendering them no longer available to replenish and fertilize floodplains, wetlands and deltas in the lower reaches. These lowland and estuarine features may then shrink in places where they once accreted. There are now unprecedented demands for water and water-supply and treatment equipment and there are increasing strains on freshwater water supplies, which in certain arid locations, are already insufficient to meet demands. In the United States, for example, between 1950 and 2000, a period when the population increased by 80%, the volume of water extracted by municipal water departments for both homes and businesses increased by 300%.[17] During that time the population grew, but people were also getting used to using larger quantities of relatively inexpensive water. More recently, since the mid-1990s, there have been the first efforts to reduce per capita consumption. Population, however, still grows. (There is more on the history of water supply and sanitation in Chapter 2. Chapter 4 describes how cities are supplied with water.)

Limits

There is a commonly encountered attitude and expectation that municipal water companies and utility companies can always find a way to bring water to where it is needed. In a certain sense that is correct – with sufficient investment, equipment, infrastructure and expenditure of money and energy it is usually possible to import freshwater or treat salty or tainted water. Even when conventional sources of water from rivers and aquifers are exhausted or unavailable, canals, tunnels and pipelines can be constructed to move water across great distances, or sea water can be made fresh in desalination plants. Foul water can be treated and recycled. There are problems, however: construction, operational and maintenance costs continue to rise and there are losses of biodiversity even in the more remote areas from where water is often abstracted. Dirty water costs much more to purify than the relatively clean water that emerges from underground aquifers or steadily trickles from forested slopes into upland lakes. The emerging problem for those wishing to rely on conventional energy-intensive approaches to supplying and treating water is that the long-term trend for energy prices has been for these to increase – a trend that is likely to continue, despite occasional dips in price like that associated with the current (but temporary) 'tight oil' boom in North America. Electricity prices vary considerably from country to country; however, broadly speaking, prices have followed those of fossil fuels, which tripled between 2000 and 2010. As well as the overall increase, during that period there was also the price spike and crash of 2008–9, which has added an element of uncertainty, making planning even more problematic. Some commentators are now convinced that cheap fuel has returned for good, but most geologists will point out that the supply of oil cannot continue indefinitely. Electricity generation itself, as well as being required to pump water, also requires water for cooling, with an estimated 15% of all water extracted from the environment being used by power stations.[18] Energy is needed to supply water and water needed to supply energy – the so-called energy-water nexus.[19]

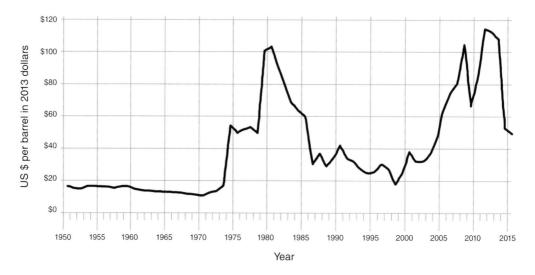

Figure 1.2 Oil prices 1950–2015.

Sanitation

The authorities usually move quickly to ensure that utility companies send sufficient quantities of drinking water to cities, even if the provision is unsatisfactory in some way – after all, we cannot live without it – however, it usually takes much longer for adequate sanitation to be provided (sometimes centuries). For example, the New River,[20] an artificial waterway designed to bring drinking water to London, was opened in 1613, but it wasn't until the 1860s, following the 'Great Stink' of 1858, more than two centuries later, that sewers were installed to divert raw sewage away from the centre of the city.[21] A few decades later, in 1900, sewage treatment finally commenced, nearly three centuries after piped supplies were installed.[22] The result of those works was that London never again suffered major outbreaks of the deadly waterborne disease of cholera. In many cities, especially in informal or squatter settlements, people still do not have access to a toilet and in the majority of cities sewage still continues to enter watercourses without any form of treatment. A third of the global urban population, about 1 billion people, has inadequate access to sanitation, resulting in the premature and avoidable deaths of more than 2 million city dwellers each year (that is equivalent to more than 5,000 people each day).[23]

Pollution

Sewage entering the wider environment is not only a threat to human health but it also causes severe damage to ecosystems. Faeces can smother aquatic fauna and, as bacteria break down the excess organic material, there is a decrease in dissolved oxygen, leading to the death of fish and most species of invertebrates. Even modest increases in dissolved nitrates and phosphates from sewage may cause eutrophication, leading to algal blooms and the subsequent dieoffs and decay, which cause those low levels of dissolved oxygen. The most badly affected watercourses can become devoid of wildlife, apart from a few pollution-tolerant mud-dwelling invertebrates like midge (Chironomid) larvae and tubifex worms. Although water quality in rivers in some of the developed countries has been improving for a few decades, following the installation of treatment plants and the closure and transfer of factories overseas, water quality in most of the world's rivers and estuaries continues to decline. The problem has spread to coastal waters with several hundred hypoxic or 'dead zones' now recorded close to many of the world's major population centres.[24] As suggested by the name, 'dead zones' no longer support fisheries. Many of these polluted coastal zones have also lost the capacity to function as spawning grounds for fish that spend part of their life cycle offshore.

Many formerly important inshore fisheries are now closed and those fishermen continuing to seek their living from fishing are often now having to sail far away from their home ports (sometimes to another hemisphere) to find fishing grounds with sufficient stocks to fill their ever larger nets and keep their floating factories occupied.

Urban Drainage

Where agricultural land has been replaced by the sealed surfaces of conurbations (mainly roofs and roads), the characteristics of surface water runoff change for the worse. Urban drainage tends to rely on pipes that discharge water into watercourses with great speed, with the water carrying with it pollutants, which in turn damage aquatic ecosystems. Urban storm water, the excess rainwater that runs off of roofs and paved areas, is typically directed through gutters and downpipes directly into underground drains. Most storm-water drains include grilles to catch litter and chambers that intercept silt and oil (but not all pollutants). Water is then directed into pipes, normally laid on a gentle gradient, ultimately sending the polluted water to a watercourse. This has the desirable effect of drying out the surface of streets rapidly but it causes large volumes of water to enter watercourses in a short period, where silt is deposited and pollutants discharged. Small watercourses suffering from such flashy flows tend to be lined with masonry or concrete to prevent erosion and, as they become more dangerous and unpleasant, over the years they tend to be covered over to become underground culverts, forgotten between maintenance visits and largely devoid of life. Quite large streams have been lost underground in this way and larger urban rivers put into deep concrete channels with trapezoidal cross-sections. A notable example of the latter is the Los Angeles River, often no more than a trickle but a watercourse that grows rapidly into a torrent following a winter storm. Storm water also causes problems in older cities with so-called combined sewers, where surface water drains are interconnected with foul sewers.[25] When rainfall is heavy, sewers leading to treatment works can be overwhelmed with rainwater, resulting in the discharge of raw sewage through overflows, which leads to sewage entering watercourses and, in severe cases, may even cause sewage to back up and flow out of the drains and into the streets. New developments do not combine foul and storm-water drainage; however, legislation designed to protect rivers from pollution means that many municipalities and utility companies in North America and Europe are investing billions of dollars and euros respectively into the retrofit of tunnels, retention basins and other projects designed to alleviate or eliminate the problems caused by combined sewer systems. Now there is increasing interest in Low Impact Development[26] or Sustainable Drainage Systems,[27] which mimic natural drainage by intercepting, detaining, attenuating and infiltrating rainwater and promoting evapotranspiration through the use of natural features, thereby keeping rainwater out of the sewers. (Read more on near-natural or sustainable drainage systems in Chapter 7.)

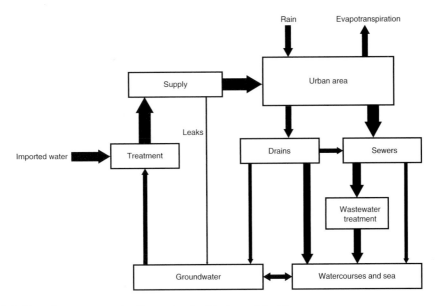

Figure 1.3 The urban water cycle. Illustration by Marianna Magklara.

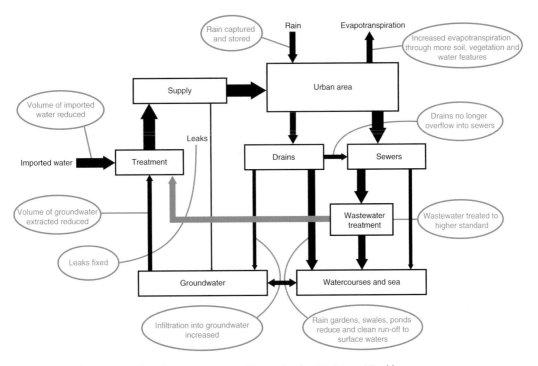

Figure 1.4 The sustainable urban water cycle. Illustration by Marianna Magklara.

Potable Water

The reliable supply of potable water that is distributed to every street, home and business in developed countries may be taken for granted by most of us who live in those places. Although taken for granted, piped supplies of potable water truly enrich the lives of those who enjoy this facility, improving health and saving time. These remarkable systems, however, have been provided at a cost to the wider environment. The assumption by consumers that there will be a constant high-quality uninterrupted supply of potable water means that, without a second thought, people install and use extra bath-rooms, showers, washing machines, power washers and irrigation systems that significantly increase consumption. Water consumption in some affluent countries, for example, like the United States and Australia, exceeds 500 litres per person per day. Compare this with a typical less-developed country, where the figure for per capita daily consumption can be as little as 50 litres.[28] Water piped into homes in developed countries is usually safe to drink, which requires expensive treatment, yet very little of the water which flows into people's homes passes their lips. In the United Kingdom, typically a third of that excep-tionally clean water is used for toilet flushing (although this figure is lower in most other Western European countries) and another quarter for showering and bathing.[29] There is also concern over the waste of water when people run taps whilst they are cleaning their teeth or washing vegetables (a running tap wastes around 6 litres every min-ute). Most of the potable water that flows into homes is subsequently wasted and, especially since the 1990s, there have been campaigns to address the problem following the realization that increasing supply may not always be affordable or feasible in the future. Residential water consumption is metered in most developed countries, however there are a few exceptions (for example, in the United Kingdom only 40% of homes have water meters installed). The metering of water is useful. It assists with the identification of leaks and has been shown to reduce consumption by 10%.[30] In many cities, concerted efforts have been made over an extended period to reduce demand for domestic water using economic incentives (for example, charging for heavy use) as well as subsidising water-saving equipment and gadgets. In addi-tion there have been many initiatives to sponsor information cam-paigns designed to change attitudes and behaviour.

Waste

Another major cause of waste in urban water supply systems is leakage within the distribution network – a problem that varies considerably in extent between cities and nations. Losses of water from within the water distribution network in Germany are less than 10%. By comparison, in Albania, losses were once reported to be as high as 75%.[31] Losses occur through corrosion, leaky joints, damaged valves and faulty hydrants and pipe breakages. Repairs are expensive – it is often perceived as cheaper to increase supplies to compensate for losses rather than to repair an underground leak; however, attitudes are changing. Most suppliers now have programmes of testing for and monitoring of leaks and are beginning to appreciate the benefits of initiating repairs at an earlier stage. Of course, broken and corroded pipework must eventually be replaced if it is to continue to perform to a satisfactory standard, that is, a standard where adequate water pressures are to be maintained. As well as affecting supplies, burst water mains can cause serious flooding to properties and may require street closures whilst repairs are made, leading to travel delays and other costly disruptions. (Read more about how demand is being reduced in Chapter 8.)

Rainwater Harvesting

Falling rain is one of the cleanest sources of water. It can be contaminated by airborne pollution but is nearly always cleaner than river water. Where rainwater is harvested it is usually collected from roofs and is therefore more likely to be close to the homes or businesses where it is required. The technologies required for harvesting rainwater are relatively simple and the technique has been used for millennia, ever since cities were first established. There are a few problems associated with rainwater harvesting – roofs can be contaminated with bird droppings and other pollutants and if water is not carefully stored it can be spoiled by algal growth or invaded by aquatic microbes (including some pathogens) and invertebrates that lay their eggs in water (notably mosquitoes and midges) – however, problems can be overcome easily through the adoption of simple safeguards. Storage tanks represent the most significant cost of rainwater harvesting (usually around 90%), although overall the cost of rainwater harvesting, when compared to the cost of securing other water sources from further afield, is relatively low. The capacity of rainwater harvesting systems is largely determined by the availability of a suitable collection area (often a roof or another clean, impermeable surface), rainfall patterns, the likely duration of dry periods between rainfall events, consumption rates and the volume provided of storage tanks. Rainwater harvesting in certain rural locations, like Gansu Province in China, for example, is well established; however, rainwater harvesting in most modern cities has been largely abandoned and forgotten since the advent of reliable water supply networks in the period following the Industrial Revolution. There is now a renewed interest in the rainwater harvesting, however. There have been thousands of schemes implemented in Germany, for example, and new associations of practising and aspiring rainwater harvesting companies are springing up all over the developed and developing world.[32] (To read more about how cities are harvesting rainwater see Chapter 9.)

Recycling

Once used, water can be captured, cleaned and recycled. The easiest source of water to exploit in this way is greywater (sometimes known as sullage), that is, water that has been recently been used for baths, showers and washing.[33] Greywater usually contains soap and detergents that can be easily screened and filtered using proven and affordable technology. As long as premises are occupied, there is a reliable and predictable stream of greywater (in contrast to rainwater), with typically a quarter of the potable supply to a building being subsequently available as greywater. The collection of grey water requires changes to the typical plumbing system, with the wastepipes from sinks, baths and showers connected to filtration systems and storage tanks. In some jurisdictions, regulations originally devised to minimize the risk of contamination of potable water by wastewater mean that the use of greywater is prohibited. In addition people in some cultures are squeamish regarding the suggestion that waste water is recycled. Water that carries sewage, so-called blackwater, can be treated to a quality whereby it is suitable for reintroduction into the potable water supply system. In most situations sewage is treated to a level whereby it is safe to discharge into rivers or the sea without an immediate threat to human health but in countries where water is becoming scarce or where there is a dependency of water imported at considerable cost from other jurisdictions, there is a strong interest in recycling sewage and treating it to a high standard. Singapore, for example, where the government wants to reduce the amount of water imported from Malaysia, is using microfiltration, reverse osmosis and ultraviolet light to turn sewage into drinking water as part of its NEWater Programme.[34] (For more on the recycling of urban water see Chapter 10.)

Biodiversity

Extinction of species is a natural process; however, conservation biologists have become alarmed by the current accelerated rate of extinction, said to be more than a thousand times the earlier 'background' rate. These extinctions are closely associated with habitat loss and overexploitation of ecosystems, which has increased in parallel with population growth during the twentieth century. The United Nations' Millennium Ecosystem Assessment, published in 2005, estimated that between 10 and 30% of all mammalian, bird and amphibian species are threatened with extinction due to human activity.[35] This loss of biodiversity is a major threat to civilization because when species are lost and genetic diversity eroded, ecosystems become degraded or may even collapse, leading in turn to the loss of the ecosystem goods and services (including food, wood, fibre, air and water purification and climate regulation) that we all depend on. The biodiversity of aquatic inland ecosystems, including rivers and lakes, has been particularly badly affected by urbanization. Wetlands have been drained to create more agriculture to feed to burgeoning urban populations. For example, 73% of the marshes of northern Greece have been drained since 1930, resulting in a decline in their associated flora and fauna, including wildfowl. Globally, more than 40% of all rivers are now intercepted by dams, which have had a severe impact on fish migration patterns and breeding success. Modifications of rivers have resulted in the loss of habitat features, which has compounded these problems, leading to the collapse of most inland fisheries. Many endemic fish species are threatened, and inland aquatic ecosystems are being lost even before biologists have had a chance to describe them. Most lowland aquatic ecosystems worldwide have been severely degraded or even destroyed in many cases.[36] Extinct species are lost forever and many of the little known and poorly understood subtleties and complexities of ecosystems can never be recovered once damage and degradation has occurred.

Restoration

River and stream restoration is possible and desirable. Reducing the volume of water abstracted from the wider environment will help to reduce our reliance on dams and will allow watercourses and associated species to recover once dams are removed. It is also possible to restore watercourses and surface water features in urban areas, helping to restore the connections between upstream and estuary and thereby making cities better places to live. The movement towards sustainable drainage systems or low impact development (as described in Chapter 7) has tended to emphasize the issue of flood management, however it constitutes a real opportunity to create wildlife habitat and restore biodiversity in the grey urban deserts and canyons that many of us inhabit. The water-sensitive city can reduce its impact on the wider environment; however, the water-sensitive city should also be a biodiverse city. Watercourses can be reinstated and naturalized for the purposes of restoring biodiversity and terrestrial features modified to intercept, infiltrate, evaporate and store water can also be habitats planted with native species in natural associations and plants known to benefit wildlife. This represents an opportunity to restore biodiversity in towns and cities, where most of humanity lives and works.

The Future

I predict that the citizens of the future water-sensitive city will see more water and will enjoy more interaction with water. Once water is brought to the surface and kept out of pipes, designers can find ways of celebrating the flows and forms of, intermittent, usually trickling, but occasionally cascading, water. This can be ornamental and playful but also practical. People, especially children, love to play with water. Water stirs our curiosity and fascinates us. It stimulates experimentation and stimulates the imagination. Playing with water encourages us to communicate and interact and is suitable for everyone regardless of age, gender, culture or ability. On a hot day, there are few activities more enjoyable than splashing around in clean, cool, clear and fresh water. Rainwater that has been carefully stored and appropriately treated can be safely used for play and outdoor water features. Having more soil and water available will also help the citizens of the water-sensitive city to grow more of their own food. Recent initiatives in France, China, the United States and even the Gaza Strip have demonstrated how urbanites can grow food on underused corners, roofs and even walls in the most difficult circumstances.[37] There are some highly technical ways of doing this, with pipework and automation; however, even a simple planted pot watered by hand can be productive and a means of escaping the worries of the world for a few moments each day. The hands-on, volunteer-led approach to urban greening may also extend from private spaces into the public realm. As local authority budgets are cut, volunteers are filling the gaps. Take, for example, the roadside rain gardens created and managed by citizens in Portland, Oregon. When bureaucrats block urban greening initiatives, guerrilla gardeners may make the plantings, if necessary, under the cover of darkness.[38] Eventually embarrassed officials relent and allow more volunteer involvement or, if that is too difficult, may then improve their own efforts at urban greening.

Privatization and Regulation

It is assumed by many of us that the provision of water and the treatment of wastewater is little more than a commercial matter, a problem to be solved by the private sector, eager to satisfy its customers by providing an excellent product or service, with government regulation ensuring that strict standards are met. The neoliberal economists who first suggested the privatization of utilities and the politicians who enthusiastically pursued this agenda, were most interested in the narrow agenda of economic growth and efficiency, lower taxation and the weakening of worker organizations, rather than more important issues of integrated catchment management, water conservation, water-quality improvements, reducing the production of greenhouses gases, tackling climate change or halting biodiversity losses. Public water supply and sanitation authorities were first privatized in the United Kingdom in 1989 and organizations like the World Bank are still enthusiastically promoting various forms of privatization, including public-private partnerships, where full privatization is perceived by local people to be an unacceptable change. The privatization of water supply and sanitation services is still controversial. Currently close to a billion people receive their water from privately owned corporations. Where they are well established, those who are ideologically committed to privatization will, no doubt, insist that the private sector will continue to play its key role in the water cycle of the future; however, it is clear that there are a number of problems with the current model. Little accountability, a lack of transparency, few opportunities for citizen participation and inflexibility and spiralling costs are common complaints. Authorities must retain an independent regulatory role and this has significant establishment and running costs. Regulation can effectively maintain high water quality and set limits on price increases; however, private utility companies will inevitably be more concerned with their own profits and asset values, rather than general concerns over scarcity, lower water quality, the drying of the wider landscape, losses of ecosystem services or biodiversity. There is also a problem with new concerns related to climate change not being addressed because these issues have usually been outside of the scope of the commercial agreements struck between corporations and governments. If private corporations do continue to provide water, drainage and sanitation, contracts may need to be rewritten to allow more flexibility to promote stronger efforts to bring about more climate change mitigation and adaptation, as well as the conservation of biodiversity and the restoration of ecosystem services.

Coordination and Cooperation

There is a growing consensus that if the security of water supply and levels of sanitation that protect both human health and restore the aquatic environment is to be achieved, more will need to be done by all of us. This requirement goes beyond the city limits or the piped network of the water supply and treatment companies. Comprehensive plans will be required that integrate water use within whole catchments. This will require new levels of cooperation between interest groups, landowners, agencies and administrations – not only within regions and nations but also, on occasion, between nations. Land-use management plans will need to be linked with river basin/ catchment management plans. The passage of water through the managed landscape will need to be modified in ways that consider water quality, biodiversity, flooding, energy and water supply. There will also need to be great efforts made to collect, recycle and reuse water in urban areas, with an increased effort to make processes more water efficient and people more committed to the conservation of water. The measurement and monitoring of water will need to be improved. Data will need to be shared and used to inform decision making, with water infrastructure systems monitored in real time as an integrated operation within the 'smart cities' of the future.[39] The 'smart cities' will need to become part of 'smart catchments'. Water management will need to become more flexible to enable us to deal with the vagaries of the weather, a problem that is predicted to become worse with climate change.

Towards a Better Future

Whatever the direction chosen by politicians, businessmen and citizens, it seems likely that the forces associated with climate change, population growth and water and energy shortages will transform cities and their hinterlands. Although these changes are unprecedented and will surely bring pain, there are also some reasons for hope. Waste and inefficiencies can be tackled and ecosystems can be restored. People are able and willing to identify strategies and techniques for creating a water-sensitive civilization. Some cities are already embracing a holistic approach to urban water management. A good examples are the Australian cities, which have promoted the concept of Water Sensitive Urban Design (WSUD), whereby the city optimizes the whole urban water cycle including stormwater, wastewater and groundwater management along with water supply, processes that are usually considered in isolation and are the responsibility of many different organizations regulated by a wide variety of different statutes and bodies.[40] (Chapter 14, the final chapter, looks at how the water-sensitive city of the future might look and function.)

2. A Brief History of Water Supply and Sanitation

Genesis

The earliest cities could only be established where freshwater was plentiful and relatively clean, usually near rivers or springs. River valleys also happen to have seasonal floods, which bring the fertile sediments that farming thrives on. The abundance and concentration of food calories created by farming and the creation of granaries to overcome the problem of seasonal scarcity meant that large settlements were possible for the first time in history. However, when people concentrate together in large numbers, the contamination of freshwater with life-threatening or debilitating pathogens becomes a serious problem.[1] Although people were not aware of the causes of disease in ancient times, they did have an innate disgust for water tainted with faeces and were driven to seek cleaner and more secure supplies of freshwater. When populations were low and cities few, it was usually possible to find adequate quantities of clean water within a short distance; however, channelling a heavy liquid like water long distances requires a means of containment (ducts or pipes) and is a time-consuming and material- and energy-intensive operation, especially if the terrain is difficult.

The Water Sensitive City, First Edition. Gary Grant.
© 2016 John Wiley & Sons, Ltd. Published 2016 by John Wiley & Sons, Ltd.

Bronze Age

Between 2600 and 1900 BC, during the Bronze Age, the Harappan civilization flourished in the Indus Valley, an area located in what is now Pakistan. In common with other civilizations of the time, its wealth was a consequence of the agricultural abundance associated with the fertility of the floodplain. One of the cities from that period was Mohenjo-daro, which housed an estimated 40,000 residents in rectilinear brick and timber dwellings. Water was supplied from over 700 cylindrical brick-lined wells, including large central wells and many more, smaller satellite wells, with each cluster serving groups of houses. It has been estimated that there was one well for every three houses. There was also a very large public bath (the Great Bath – a prominent feature with a religious purpose perhaps) and bathing platforms. The sewerage system was remarkable for its time and nothing comparable would be seen until the Roman period, nearly 2000 years later. Although there was some use of simple soak pits and sewage-collection tanks, waste left many of the individual dwellings either directly down a chute or through clay pipes into brick-lined channels, which then ran along the streets, with the flows ultimately discharging into the river. A typical U-shaped sewer would run along a street and be up to 250 mm wide and 500 mm deep and protected by removable timber covers or flagstones. The gradient of the drains was typically 1 in 50 (modern drainage pipe gradients are usually between 1 in 40 and 1 in 110). The drains were often curved and cesspits were installed in congested sections, presumably to allow settling of solids and easier cleaning.[2] It has been suggested that Mohenjo-daro was destroyed by an exceptionally severe flood, something that is always a possibility for settlements located on a floodplain.

The First Aqueducts

Also around 4000 years ago, aqueducts began to appear in Mesopotamia and Minoan Crete. These structures meant that water could be brought from distant springs or streams, increasing the reliability and volume of supply and usually bringing water of a higher quality. Aqueducts meant that crops could be irrigated and therefore grown in new locations, reducing the reliance on river flooding. Aqueducts also meant that people could begin to enjoy regular baths – important for hygiene, of course, but also an activity with great social and religious significance. Water supply infrastructure of this kind is vulnerable to attack or sabotage, which suggests that these features could only be built and maintained once sufficiently large territories could be secured.

Nineveh

By 850 BC the city of Nineveh in Babylonia, controlled by the Assyrian emperor Sennacherid, had large palaces and a population of more than 100,000. It was therefore a settlement that must have required a sophisticated water supply system. It had already been an important city for more than 2000 years, but had reached new levels of grandeur with, for example, monumental sculptures made from stone brought from quarries more than 50 km away. A series of canals brought water from the hills, a system that included a section of aqueduct, which was about 65 km away from the city at Jerwan.[3] This unique and wonderful aqueduct was 9 m high and carried a canal, 22 m wide and spanning 280 m of a river valley. It has been estimated that more than 2 million stone blocks were used to make the arches of the aqueduct. The ruins remain; however, it seems likely that a river, which passes under the aqueduct, may have undermined the structure at some point in the past. More recently it has been suggested that this water supply system also supplied the legendary Hanging Gardens, which may have been located on a mound close to the city of Nineveh and not in Babylon as earlier writers had assumed.[4]

The Nile

From ancient times, in a typical year, from July to September, the River Nile would break out of its banks to deposit fertile sediment onto the floodplain. That rich alluvial soil is the basis of the agricultural wealth upon which the ancient civilization of Egypt was built. From around 3000 BC, a leader by the name of Menes united the kingdoms of Egypt and founded, perhaps, a new capital at Thinis. The details from that early period are uncertain, however it seems to mark the beginning of the construction of water control systems, including canals, irrigation ditches, sluices and basins – the essential agricultural infrastructure for a civilization that was to last for a further 3000 years. For drinking water, the Egyptians relied on wells. A typical well would be around 2 m across, round or square in plan and lined with limestone masonry. Excavations of a well dating from the period of Rameses II (1279–1213 BC) found that the bottom of the well had been filled with layers of broken pottery, more than 1 m in depth, and it has been suggested that these shard layers were designed to filter sediment. This technique is similar to the use of pebbles in wells built today. It does not significantly improve the quality of the water in terms of microbes, which are unaffected, but the water is much clearer and inviting. Another interesting feature of the upper section of that excavated well is the impermeable clay layer on the outside of the masonry, which protects it from adjacent, possibly contaminated, surface water. The location of this particular well appears to have been carefully selected and is likely to be representative of many others. It was sunk into a lens of sand, which sits above the general alluvium, a geological feature which stays waterlogged, even when flows in the Nile are low.[5]

The Minoans

At the height of the Minoan civilization, sometime between 2200 and 1700 BC, a great earthquake occurred, which caused abandonment and ruin but also brought about the preservation of some interesting features.[6] Located in modern Crete, the most important Minoan building complex was the Palace of Knossos, which consisted of more than 1300 rooms (this was the original Labyrinth of legend). The settlement was constructed from sundried mud bricks set upon limestone blocks. Buildings had high thresholds, reducing their vulnerability to flooding and making underground pipework and heating easier to install. The Minoans were the first people to use underground pipework for sanitation and water supply. They used tapered clay pipes, typically 110–150 mm in diameter, with overlapping joints of about 75 mm. The tapering created a jetting action, which may have helped to clear sediment. Aqueducts brought freshwater from the source of the Kairatos River at Archanes; however, they also collected rainwater from roofs. Knossos was also the site of Europe's first flushable toilets, using rainwater held in cisterns and directed through conduits built into the walls. As well as waste pipes, there were also larger open sewers designed to deal with the heavier flows experienced during winter storms.

Qanats

In the early part of the first millennium BC, people living on the Iranian plateau were excavating, by hand, sloping underground conduits to take water from mountains to agricultural areas in the valleys. These qanats, as they are known locally, are typically half a metre wide and a metre high and vary in length between 1 and 70 km, delivering up to 500 litres of water per second. There are still some 22,000 qanats in Iran, continuing to make life possible in some very arid places. The technique spread to North Africa and Spain in the west, Arabia in the south and India to the east. The method of construction involves the sinking of a series of vertical shafts, which are then connected by a gently sloping underground conduit. Moving water underground reduces evaporative losses and is relatively unaffected by pollution and the animal vectors that spread disease. Overabstraction tends not

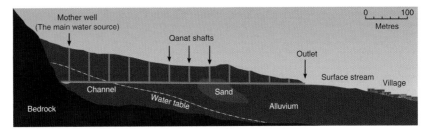

Figure 2.1 A typical qanat. Illustration by Marianna Magklara.

to occur because pumping does not take place – only water that is readily available from the mountain aquifer is transported by gravity. Closely associated with the qanats was the system of 'jar irrigation', whereby narrow-necked permeable clay jars were filled with water and buried close to plants that required irrigation. This method of irrigation, which encourages subsurface percolation, reduced evaporative losses. Until 1955 (when a large reservoir was created and piped water system was introduced), water was supplied to Tehran's 1.5 million residents by 36 qanats, some of them more than 250 years old.[7]

Pompeii

The eruption of Vesuvius in 79 AD covered Pompeii with volcanic pumice and ash and preserved intact features usually missing from archaeological remains, including most of the water supply and drainage systems. In the beginning, the people of Pompeii collected rainwater in a shallow pool within their homes (known as an impluvium) or had private wells. By 20 BC, however, the Serino Aqueduct was constructed to bring water from the mountains to eight towns in the Bay of Naples. At 140 km in length, it was one of the longest aqueducts of antiquity and included a total of 2 km of tunnels. Water was directed towards Pompeii on a spur from the Serino Aqueduct and arrived at the Castellum Aquae, a large building at the highest point of the town. This structure was the point from which the water supply for the town was controlled.[8] There were three outlets from the building. Each could be regulated through the use of a gate, and water was directed through lead or ceramic pipes to public baths, the houses of wealthy citizens, or public fountains. Having three outlets meant that there was the option of cutting off supplies to the baths or dwellings during periods of low flow. Although many dwellings had private cesspits, where waste could dissipate into the permeable ground, in general wastewater was directed towards the paved roads, which acted as open sewers. Pavements (pedestrian sidewalks) were elevated and kerbs channelled the flow of filth along the roadways. Stepping stones allowed pedestrians to cross the street without stepping in the roadway and strategically placed gaps between the stones allowed wheeled vehicles to pass unhindered.

Byzantium

In AD 330, the Emperor Constantine made Byzantium (which became Constantinople and later Istanbul) the new capital of the Roman Empire. The new city was vulnerable to water shortages, which led to the planning of an impressive supply system. By AD 373, after 30 years of construction, the Emperor Valens was able to witness the commencement of operation of the new aqueduct, which brought freshwater 150 km from the Istranja Mountains near Vize. Over the ensuing 150 years a further 400 km of channels and other features were added, including masonry bridges, tunnels, cisterns and open-air reservoirs, with the longest single stretch 250 km in length. The system, which is probably the most extensive from antiquity, continued to supply up to 500,000 people for around 800 years, into the beginning of the Ottoman period. Only the cumulative impact of earthquakes put the system beyond repair.[9] The original Valens aqueduct was around 600 mm wide and 1 m high and was supplemented by larger parallel channels in later centuries. Once the water had been delivered to the city limits, it was distributed to around 100 cisterns within the city walls, ranging from small private tanks to large open-air reservoirs on high ground in the suburbs, altogether storing up to 900,000 m³ of water. As well as a storage function, some of these cisterns were also control points.[10] By bringing water from reliable and plentiful sources in the mountains of Thrace, the people of Constantinople were able to overcome the annual shortage of summer rainfall and the absence of local ground water during periods of drought.

Yucatan

Much of the Yucatan peninsula, in the southeastern corner of modern Mexico, is underlain with a stratum of limestone, which constitutes a large underground aquifer. Tropical downpours soon disappear into the porous limestone and there are few surface streams in this region. The Mayan civilization, which reached its peak during the period between AD 250 and AD 900, relied on more than 2000 natural underground caves (known locally as cenotes) and man-made underground cisterns (known as chultuns) for its water supply. These cenotes and chultuns were supplemented by many more lined surface depressions (known as aquados), which were created when building materials were quarried. A typical chultun would be 5 m deep, rendered with lime mortar to prevent leakage and able to store 1500 litres of rainwater runoff from nearby buildings or plazas – enough to supply 25 people.[11] Despite the ingenuity of the water supply engineers of that time, the classic period of Mayan civilization ended around AD 900, possibly because of a relatively mild yet prolonged drought. The annual water deficit continued for several decades, which meant that they were unable to replenish their underground reservoirs, which eventually ran dry.[12]

The Incas

El Nino refers to the warmer sea-surface temperatures that develop periodically off the Pacific coast of South America. During these periods, which return every few years and which can last for a few months and occasionally as long as 2 years, warm air brings heavy rains to the deserts of northern Peru. The irregular and unpredictable patterns of rainfall experienced in this area (as well as the other deserts in South America) meant that there is a long history of efforts to create dependable water supplies, both for irrigation and potable use. Tipon, in Peru, was an Inca settlement from around AD 1200 to AD 1534. It was a walled estate, covering about 200 hectares in the Huatanay River valley. There remains a series of agricultural terraces originally created by infilling the valley with stones. The flow of water from one terrace to another is precisely controlled by these stone structures. Excess surface water is directed along a 60 m long aqueduct and there is a reservoir; however, what is particularly interesting and unusual about this site is the storage and use of subsurface flows, increasing the volume of water stored overall and making the system able to function for longer during periods of drought.[13]

Figure 2.2 Tipon. Photograph by Jason Westlake.

Qi

The encyclopaedia known as Guanzi, which was compiled during the fourth century BC in the northern Chinese state of Qi, provided the following advice on urban planning: 'The capital should be built either at the foot of a great mountain or near a grand river. Stay away from dry lands at high elevation in order to get enough water and stay away from water at low elevation to prevent flood and reduce the need for drainage and embankments.' It is clear from this advice as well as the operation of various projects, some of which survive to the present day, that the management of water was undertaken with considerable sophistication in ancient China. An example of one of those projects is the Dujiangyan irrigation system, located on the Min River in Sichuan Province, which was first established in 256 BC. This system is still in use and is a UNESCO World Heritage Site.[14] The Dujiangyan irrigation system uses a river weir and bypass channel. The combination of weir and bypass channel diverts water in a reliable way in order to irrigate hundreds of thousands of hectares of farmland. It does not involve the use of a dam and therefore does not interfere with the migration of fish, nor does the arrangement cause major siltation (which is a problem with dams).

Lijiang

Another World Heritage Site in China is Lijiang in Yunnan Province, which was established as a new town in the thirteenth century AD.[15] A main water supply channel branched into three and then into a network of stone culverts and channels, which supplied every dwelling. This necessitated the construction of a large number of bridges (354 in all) and it is these bridges, which gave the town its name. The town is flanked by two substantial watercourses – the Western River and Central River – and this presented an opportunity for engineers to exploit the difference in levels between these two watercourses to create a separate drainage system which, in effect, flushed waste from the streets. This flushing arrangement was probably a unique form of municipal street cleaning and sanitation.

Medieval and Early Modern Europe

In Europe, during the Medieval period, the streets were open sewers, often clogged with the waste from butchery and other food-preparation processes. There was no running water (as there had been in ancient Roman cities) and the accumulation of decaying foodstuffs and faeces meant that deadly contamination of wells and drinking water must have been a frequent occurrence. Dysentery (along with many other diseases) was commonplace and life expectancy low by modern standards, for poor and rich alike. For example, King Henry V of England probably died of dysentery (a disease usually caused by pathogens in contaminated water) in 1422 at the age of 35.

Early Victorian Period

Little progress was made with the improvement of sanitation until the early nineteenth century, when experiments with sand filters led to the establishment of the world's first treated public water supply, provided by the Chelsea Waterworks Company in London in 1829. The Chelsea Waterworks Company had originally been established in 1723 and along with a number of other similar private water supply companies in the capital, took advantage of the steam engine to pump water from canals and rivers to service reservoirs. The Metropolitan Water Act of 1852 introduced legislation to regulate the new water-supply companies, banning the extraction of water from the tidal Thames and requiring that water be 'effectually filtered'.

Germ Theory

During the early Victorian period, however, the way that water harbours and carries pathogens was yet to be unveiled and the prevailing belief was in the *miasma* theory, which held that disease was spread by noxious air. This belief began to change in 1854, following the physician John Snow's meticulous mapping of individual cases of an outbreak of cholera in the Broad Street neighbourhood of Soho, London. Snow identified a particular water pump as the source of the outbreak. A subsequent investigation found that a leaking cesspit was contaminating well water.[16] It is also interesting to note how advances were being made in medical care at the time, with the pioneering work of Mary Seacole[17] and Florence Nightingale,[18] who were systematically improving the treatment of soldiers injured during the Crimean War. As the second half of the nineteenth century unfolded, the germ theory of disease established itself. Pioneers in the field included Ignaz Semmelweis, who advocated hand washing in the Vienna General Hospital, Louis Pasteur, whose experiments proved beyond doubt that germs spoiled food and drink and thereby caused disease and Robert Koch, who isolated the species of bacteria which cause tuberculosis, cholera and anthrax.[19] The work of these pioneers and the application of this new knowledge, with, for example, the use of disinfectants in a clinical setting, was followed by the treatment of water supplies. Chlorination as a way of sterilizing drinking water was pioneered in Germany and England in the 1890s and was becoming mainstream in industrialized countries by the early part of the twentieth century.

The Great Stink

In the summer of 1858, just a few a few years after those first insights were made into how disease could be spread by water supplies, came the 'Great Stink'. The River Thames in London had become so polluted with sewage and other food waste that the smell had become unbearable – most notably by the parliamentarians, who even considered the possibility of moving Parliament out of the capital. The population of London was around 2 million people at the time and an estimated 400,000 tonnes of sewage flowed into the river every day.[20] Recent improvements in water supply and the subsequent more widespread use of flush toilets had caused many old cesspools to overflow and send ever more waste into both the streets and the river itself. Members of Parliament eventually paved the way for the London authorities to copy Paris (where improvements to the drainage system were already underway) and implement some long-standing plans to build a system of underground drains to intercept sewage and take it downstream and beyond the main conurbation. Problems with the contamination of water supply continued, however. For example polluted water from the River Lea entered water supply reservoirs in east London, causing the capital's final epidemic of cholera in 1866, when 5596 lives were lost. (Cholera had been first reported in England in 1831 and continued to cast its shadow for the next three decades). By the early twentieth century, cholera was all but eradicated from Western Europe, but pandemics, which took hundreds of thousands of lives, continued to affect places elsewhere in the world where sanitation was, and still is in many cases, largely absent. Outbreaks of cholera continue to occur in the twenty-first century, in Africa, parts of Asia and a few locations in around the Caribbean (mostly notably Haiti).

Modern Sewers and Sewage Treatment

London was not the first city to have a comprehensive sewer system. Hamburg in Germany was the first, designed by the British engineer William Lindley after Hamburg's great fire in 1842.[21] Working with his sons, Lindley went on to design sewers in 30 other cities all across continental Europe. In the United States, Chicago and Brooklyn were the first cities to build sewers, in the late 1850s.[22] From 1865, London's sewage was being transferred via the Bazalgette interception sewers to outfalls to the east (at Beckton on the northern bank and Crossness, south of the river). Sewage was being taken away from main centres of population by these new sewer systems, however the practice of treating sewage did not begin for another two decades or more, not until the effectiveness of 'sewage farms' (sewage treatment works) had been demonstrated by pioneers, like the chemist Edward Frankland.[23] The first sewage-treatment plant was established in Frankfurt am Main in Germany in 1887.[24] Early sewage treatment plants separated out the sludge and passed the remaining effluent through filter beds of gravel, where burgeoning populations of microbes decompose the organic matter in the waste. The technique was refined at the Lawrence Experiment Station (now known as the Senator William X. Wall Experiment Station), which was established by the Massachusetts State Board of Health in 1886.[25]

Sewage Treatment Refined

By 1912 the eighth report of the British Royal Commission on Sewage Disposal had established the principles of primary and secondary treatment stages for sewage and set a quality standard for treated effluent being discharged into the environment. Primary treatment removes floating and suspended solids (sludge) and secondary treatment is the biological process, which ultimately removes most dissolved organic pollutants. The so-called 20/30 standard, which dates from that time and which was used in most jurisdictions until at least the 1970s, refers to 20 mg/l biological oxygen demand (BOD) and 30 mg/l suspended solids (SS).[26] Biological oxygen demand refers to the amount of dissolved oxygen required by aerobic microbes to break down organic material at 20 °C over a period of 5 days and is a widely used measurement of organic pollution in water. The term 'suspended solids' refers to small particles of solid material suspended in water.

Standards for Sewage Treatment

The establishment of standards for sewage treatment at the turn of the twentieth century was soon followed by the invention of the activated sludge process. In 1912 Gilbert Fowler of the University of Manchester in England observed various experiments involving the aeration of sewage at the Lawrence Experiment Station in Massachusetts. Fowler then returned to Manchester to work with his colleagues Edward Ardern and William Lockett to create a practical method of speeding up the process whereby sewage sludge is broken down, a technique that was to rapidly spread throughout the developed world after the First World War.[27] Activated sludge treatment involves the introduction of air (with oxygen the gas required by the aerobic microbes), by using paddles or pumps, thereby creating a frothy mixture of sewage sludge and wastewater. This process creates loosely aggregated clumps of bacteria, protozoa, rotifers and other microbes, which are then held in configurations where decomposition is accelerated, making the treatment process more efficient.

Birmingham Corporation Water Act 1892

As the twentieth century began and urban populations continued to grow, industrial cities began to look further afield for their water supply. Birmingham, a large city in the industrial West Midlands in England, began to consider the possibility of sourcing water from the mountains of Wales, 110 km (70 miles) to the west. Following a campaign led by Joseph Chamberlain, the leader of Birmingham City Council, the Birmingham Corporation Water Act of 1892 was passed. It allowed the city to acquire the catchment of the Elan Valleys by compulsory purchase and to construct a series of reservoirs and an aqueduct to bring water to the city's water works. This particular area in the Welsh mountains had been selected because of the high annual rainfall, the impermeable bedrock and the steep and relatively narrow valleys, which would make dam construction easier. Pumping would be unnecessary because the new reservoirs, once built, would be more than 50 m higher than the City of Birmingham. The scheme, with two 42 inch (1.07 m) diameter iron pipes, began to supply water in 1906. The project was extended in 1961, with more reservoirs and two more 60 inch (1.52 m) diameter pipes. Further expansion of the scheme was mooted in the 1970s; however, this second extension was eventually abandoned as industry began to decline in Birmingham and concerns were raised in Wales regarding the flooding of even more valleys to create yet more reservoirs.[28]

Los Angeles and the Owens Valley

There had been initial opposition to the Elan Valleys scheme in Wales but that opposition did not reach the levels of conflict experienced in California in the 1920s. Self-taught, Irish born, chief municipal engineer for the City of Los Angeles, William Mulholland, had overseen the construction of the 233 mile-long (376 km) Los Angeles Aqueduct, which brought water from the Owens Valley, high in the Eastern Sierra mountains. The city had secured the rights to the water in an underhand way but it was the drying out of the Owens Valley that eventually provoked the farmers and ranchers to turn to violence. The ranchers were already struggling to secure enough water for their own purposes and the supply of water for Los Angeles, which led to the drying out of

Figure 2.3 Owens Valley. Photograph by Ray DeLea.

the valley to the extent that threatened farming operations. Owens Lake finally dried up in 1924. Then local saboteurs, antagonized by provocative statements made by Mulholland in newspaper articles, used dynamite to damage the aqueduct, which temporarily diverted water back into the Owens River.[29] The conflict continued for a few years, but by 1928 the City of Los Angeles has acquired 90% of the water rights and agriculture was effectively finished in the Owens Valley. Desertification was about to set in.[30] The increase in water supply to Los Angeles enabled that city to continue to grow from half a million people in Mulholland's time to nearly 4 million now; however, the problems continue, with unprecedented drought being experienced by Californians. No rain fell on Los Angeles in January 2014 – the fifth time this had occurred since records began in 1978.[31] This conflict is not unique. Conflicts triggered by water scarcity continue and are predicted to become more common with the increase in population. Per capita demand for water has increased as living standards have improved and the effects of climate change are being felt. (The demand for water is the subject of the next chapter.)

3. Demand

Basic Needs

Taking into account the requirements of a 'lactating woman engaged in moderate physical activity', the World Health Organization (WHO), according to the Sphere Standards of 2004, advises that a minimum of 7.5 litres of clean water is required by every person each day. Taking into account basic hygiene and food-preparation needs, the WHO also advises that a total of 20 litres per day is assigned for each person for planning purposes. These totals do not include water required for laundry and bathing.[1,2] The most basic need, of course, is for drinking water. For drinking alone, a person needs between 2.5 and 3 litres of water per day, depending on their own physiology and level of activity and the local climate. In warm weather people perspire more and need more to drink. Basic hygiene regimes vary from culture to culture; however, the minimum requirement for such purposes is likely to be between 2 and 6 litres per day. For cooking, between 3 and 6 litres per person per day is budgeted for by emergency planners. Beyond personal needs, further supplies of water are required for cleaning the home, watering vegetable gardens and domestic animals and for sanitation and waste disposal. Taking these other uses into account, it would seem that around 70 litres per person per day is probably as low a demand that can be realistically aimed for, although it is worth noting that not all of these activities (watering gardens or flushing toilets for example) require water of a quality suitable for drinking.

The Water Sensitive City, First Edition. Gary Grant.
© 2016 John Wiley & Sons, Ltd. Published 2016 by John Wiley & Sons, Ltd.

Personal Consumption

If 70 litres of water per person per day is the lowest basic level of demand as determined by the various requirements identified by the WHO, what are the actual levels of consumption observed in various societies around the world? In the United States, estimates of average consumption vary between 300 and 380 litres per person per day.[3] This is the volume of water consumed at home and does not include the water used to produce all the food, energy and products consumed – the so called embedded water – an issue that is discussed in more detail below. Typically, between a quarter and a third of household water in the United States is used to flush toilets, a fifth goes through washing machines and another fifth is used for showers and baths. Typically, up to a quarter of household water in the United States may be used to flush toilets, up to a fifth may go through washing machines, a sixth might be used for showers and perhaps a tenth used for in the kitchen for cooking and drinking. Up to a third might be used for watering the garden and washing cars. Leaks may also account for more than a tenth of the water consumed.[4]

In the United Kingdom, citizens consume on average about 150 litres per person per day and this has been growing at the rate of 1% per

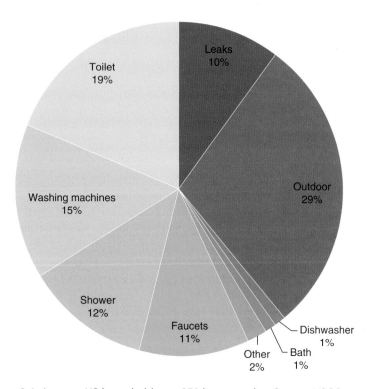

Figure 3.1 Average US household uses 850 litres per day. *Source*: USGS.

annum since the 1930s, when the rate of consumption was 126 litres per person per day.[5] Even then, the amount of water used had already risen rapidly from an estimated 18 litres per day in the 1830s, following the widespread provision of piped supplies.[6] In the present day, about 30% of water is used for toilet flushing, with about a third used for personal hygiene, in showers and baths and 13% used for washing clothes. Germany has one of the lowest domestic per capita consumption rates in the developed world. In 1990, Germans consumed around 145 litres of water per person per day (a slightly smaller quantity than that consumed by United Kingdom citizens in 2014); however, consumption has now been reduced to around 120 litres per person per day and is as low as 110 litres per person per day in a few cities, including Hamburg.[7] The proportions of water used for various purposes in Germany are similar to those used in other developed countries. The reduction in consumption in Germany since the mid-1980s has been attributed to metering in all houses and apartments, installation of new more efficient fittings and a series of campaigns designed to increase awareness of the importance of water conservation. In addition, in Germany, losses from the distribution network through leaks are the lowest in the world, around 7%, compared with more than 25% in France and Italy, for example.[8]

Water Footprint

Water consumed at home, sometimes described as the 'direct water footprint', is only part of our overall water footprint and represents less than 10% of all water consumed in developed economies. The 'indirect water footprint' is much larger. This is the water that is consumed or polluted to provide all the goods and services that we buy. Like domestic consumption, the indirect consumption of water varies considerably from country to country and even within the borders of some of the larger nations there is great variability. An added complication is that nations report how much water is withdrawn and this can be presented as a per capita figure; however, many foodstuffs and products are exported, constituting the so-called export of virtual water. Goods and foodstuffs may be exported, re-exported and imported, resulting in a complicated virtual-water budget for each nation or administrative district, with much of this virtual water consumed within large cities.[9] The water footprint may itself be divided into blue, green and grey-water components. The blue-water footprint refers to the volume of ground water or surface water incorporated into a product or evaporated during its production. The green water footprint refers to the volume of rainwater consumed in the same way and the grey water footprint refers to the volume of freshwater required to assimilate pollutants according to agreed water quality standards.

Dependency

Nations may be considered to be self-sufficient in water if they are able to provide all the water for domestic demands of good and services (however this may not be sustainable, of course, if groundwater is being drawn down or water bodies are being polluted at a rate that exceeds the rate of replenishment or recovery). Given the global system of trade, many rich nations have become dependent on others, importing water via rivers or pipelines or, more usually, goods and services with a virtual water footprint that exceeds their own water-supply capacity.

China

The national average water footprint in China (from 1996 to 2005) was 1071 m^3 per person per year, equivalent to 2934 litres per person per day. (The average daily domestic per capita consumption in China is around 80 litres, although in cities this figure is much higher and is closer to levels of use more typical of developed countries.)[10] Only 7% of the national average water-footprint figure is virtual water imported from outside of China, which means that China is close to being self-sufficient in water. However the picture is complicated by regional differences, with the north of China facing water shortages and the south of China receiving much of its food (with its associated virtual water) from the north. This has led to the proposed, and very controversial, $62 billion South-North Water Transfer Project, which aims to channel water from the Yangtze River, in the south, to the arid Yellow River basin in the north.[11] China has experienced rapid economic growth since Deng Xiaoping's social and economic reforms of the 1980s and its industry consumes and pollutes water as part of its effort to export so many products – an export of virtual water that is contributing to the domestic water-shortage difficulties that China is now facing. To give an indication of the scale of the problem, China's own government has reported that around 20% of all farmland in China is contaminated by polluted water.[12]

Germany

On average each German citizen consumes 1426 m^3 of water per person per year, the equivalent of 3900 litres per day. As discussed previously, only 120 litres of water per person per day of this is for domestic consumption. Around half of the water consumed in Germany, about 740 m^3 per person per annum, can be attributed to the indirect or virtual consumption of water associated with the production of imported agricultural products. These products include coffee, cocoa, oilseed, cotton, pork, soya, beef, milk, nuts and sunflower oil, imported from (in descending order) Brazil, Ivory Coast, France, the Netherlands, the United States, Indonesia, Ghana, India, Turkey and Denmark.[13] Although Germany pursues a strategy of reducing water consumption at home, its relative wealth and therefore ability to import much of what it consumes, drives up its indirect or virtual water footprint, in terms of water consumed in the many other countries that supply produce.

India

Some of the imbalances associated with virtual water observed internationally also occur within the single nation. India is a good example of this. In contrast with Germany, only 2% of India's water footprint lies outside of the country. Indians consume $980\,m^3$ of water per person per year.[14] The country has a large and expanding population (1.3 billion people in 2014) and there are efforts being made to increase domestic food production by making more water available for irrigation. Like China, India wants to transfer water between catchments. One such plan, known as the National River Linking Programme, is to (eventually) move 178 billion cubic metres of water per year along 12,500 km of new canals, a project that will cost in excess of $120 billion. There is already an existing trade in virtual water between states, often through the seemingly perverse movement of food from dry regions to water-rich regions, a phenomenon driven not by considerations of water resource management but by economic factors.

Indonesia

Indonesia's water footprint is $1317\,m^3$ of water per person per year.[15] Like India, Indonesia is also a nation with a rapidly expanding population, however it is also a nation of islands, without the option that India has of creating a network of water supply canals to connect different regions. Virtual water flows in Indonesia are largely characterized by the trade in rice, much of which is imported into the heavily populated and water stressed island of Java from places like Kalimantan (the Indonesian part of the island of Borneo). This arrangement has not been without its problems – the Mega Rice Project, for example, was launched in Central Kalimantan in 1996 to turn one million hectares of virgin swamp forest into rice paddies. This project involved what became a disastrous effort to clear natural vegetation and the creation of canals. The project was eventually abandoned after the area remained too dry for cultivation, even during the rainy season, leading to the proliferation of peat fires.[16]

Spain

Spain has one of the largest water footprints in the world, with a per capita consumption equivalent to 2300 m^3 per person per year (6300 litres per person per day), exceeded only by the United States and Italy.[17] Spain is also the most arid country in the European Union, with many cities in central and southern Spain receiving less than 500 mm of rainfall per year. Direct urban supply represents about 5% of total water used, the industrial sector uses about 15% (half of which is said to correspond with imported virtual water) with the remaining 80% of water consumed used by the agricultural sector. About one-third of the water used by the agricultural sector is virtual water, associated with imports, leaving the remaining two-thirds (or 53% of the total water footprint of the country) consumed by indigenous agriculture. Spain produces a wide range of crops including olives and grapes, which are well adapted to the climate and are water efficient, but it also uses large areas to produce thirstier crops like grains, vegetables and pulses, much of which is irrigated using groundwater. There is widespread concern that Spain is producing large quantities of low-value, water-intensive crops, with 60% of the water consumed by agriculture producing only 5% of the total economic value of the agricultural sector's output. Nevertheless it is believed that Spain still may have enough water for its needs, provided that improvements in efficiency can be made, along with new agricultural policies that favour water-wise crop selections.

United Kingdom

Consumers in the United Kingdom have a water footprint of 4645 litres per day (1685 m^3 per person per year), considerably more than the German consumption of 3900 litres per day. Sixty-two per cent of the United Kingdom's water footprint is indirect consumption of clothing and food from abroad. As is the case with other rich countries, these imports have an impact on the quantity and quality of surface waters and aquifers across the globe. A reduction in consumption is one possible approach to reducing this impact; however, as long as global trade continues, it will be important for the United Kingdom government, corporations and consumers to work with their counterparts in countries where this virtual water originates in order to improve agricultural efficiency and water management.

Water Footprint of Products

Demand for water is also related to the various individual products and foodstuffs that are consumed. The energy sector also consumes large quantities of water. Understanding how much water is used to produce and dispose of various products is likely to become a more important aspect of the world's water-conservation effort. We can choose to buy food and products that consume less water and we can also highlight the waste of water in particular industries so that their processes can be scrutinized and made to be more efficient. In order to ascertain the water footprint of a particular product, the amount of freshwater consumed during each step in the production process is added to produce a total figure. The Water Footprint Network, an NGO with a global outlook, which is based in the Netherlands, maintains a database of the water footprint of crops, animal products and industrial products. A few examples are described here to illustrate how surprisingly large quantities of water are used in the agricultural and industrial sectors.[18]

Meat

As societies become richer, their consumption of meat increases. In the United States, for example, about 3% of all water extracted is used to water livestock. This has serious consequences in terms of land use change, habitat loss and energy consumption as well as increasing the overall demand for water. Consumption rates do vary from country to country and between different systems of husbandry; however, the global average figures provided by the Water Footprint Network give us an indication of the scale of the problem. Meat from beef cattle is the most demanding, requiring 15,415 litres of water for each kilogram produced, compared with 87,631 litres/kg for meat from sheep and goats and 4325 litres/kg for chicken; 98% of the water included in these figures is used to produce the feedstock for the animals, with the rest required for drinking and husbandry. To put this in perspective, 1 kg of beef requires the equivalent volume of water to 100 days of domestic consumption of water for one person in the United Kingdom. The beef sector alone uses one-third of the global animal water footprint of 2422 billion cubic metres (Gm^3) of water per year; 87.2% of this water is 'green' (rainwater), 6.2% 'blue' (rivers, lakes and aquifers) and 6.6% 'grey' water (recycled).

Vegetable Crops

Looking down the food chain at vegetables, the quantities of water used are relatively much smaller than those associated with meat production, but still surprisingly large in some cases. In the United States, approximately 80% of freshwater consumed is used for irrigating crops. Sugar requires the least water; around 200 litres of water for every kilogram produced, with vegetables requiring 322 l/kg, cereals 1644 litres/kg and nuts 9063 litres/kg. As well as nuts, other crops with a relatively high water footprint are coffee, tea, cocoa, tobacco, spices,

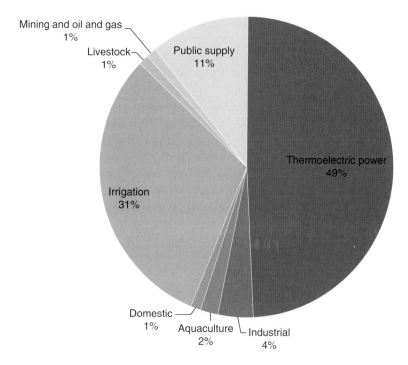

Figure 3.2 Water use by sectors in the United States 2005. *Source*: USGS.[19]

rubber and fibres. The loss of large quantities of water through evapotranspiration in regions that receive high rainfall, however, is less of a concern than those crops that consume particularly large quantities of blue water (that is water abstracted from water courses, lakes and aquifers) in relatively dry regions. Wheat and rice, for example, accounts for 45% of the global blue water footprint, an amount which has increased sevenfold during the last century.[20] The cultivation of crops in semidesert conditions using water pumped from aquifers is a special concern. The Ogallala Aquifer, for example, is used to irrigate the croplands and rangelands in the former grassland prairies of Colorado, Kansas, Oklahoma, Nebraska, New Mexico, Texas, South Dakota and Wyoming.[21] The extraction of groundwater to irrigate wheat and cotton, amongst other crops, is certainly unsustainable. The rate of extraction exceeds natural recharge rates, with fears that the aquifer could be used up within 15 years, after only 80 years of exploitation.[22]

Power Plants

Other major consumers of water are thermoelectric power plants, where steam turbines (normally driven by coal, oil, gas or nuclear fission) generate electricity. In the United States, about 3% of all freshwater consumption is for use in thermoelectric power plants.[23] Relatively cold water is required to return the steam back to water before it can be recirculated through the turbines. Single-pass cooling systems use large volumes of water, which is immediately discharged back into an adjacent river or large water body, making it available for further use. This accounts for the huge difference between the amount of water withdrawn for power plants (49% of the total - see figure 3.2) and the small amount actually consumed (lost through evaporation). Although water discharged into rivers from power plants may be relatively clean, the temperature of the water is elevated in a way that decreases dissolved oxygen and this has a negative impact on river ecosystems. Another method of using water in power plants is to recirculate water through cooling towers. Smaller volumes of water are required because water is recirculated through the system, however some is lost through evaporation and this must be replaced. A typical single-pass (or once-through) power plant withdraws 100,000 litres of water from a watercourse for each megawatt hour of electricity produced. A recirculating system is much more efficient, needing just 5,000 litres of water for each megawatt hour of electricity produced, which is the water required to replace evaporative losses.[24]

Steel

Other industrial processes also require large quantities of water, on a similar scale to power generation. The production of steel, for example, can require inputs of up to $148\,m^3$ of water per tonne of steel.[25] More typically, though, the intake of water is $30\,m^3$ of water per tonne of steel, with most of it recovered, leaving a net consumption of between 1 and $5\,m^3$ of water per tonne of steel.[26] Water is used predominantly for quenching and cooling in furnaces and mills, where temperatures may reach $2400\,°C$, but also as a solvent for cleaning. Water is also used in finishing steel, in galvanizing and plating for example. Water is essential for steel production and the availability of sufficient quantities of water of satisfactory quality is a factor in determining suitable locations for steel production plants. Water quality standards are demanding, however, in most jurisdictions. The way that steel is produced and the exact role of water in the process does vary considerably, even though water consumption rates in the steel industry have declined by more than 50% from their peak in the 1970s, there is still considerable scope for further reductions in use.[27] Although water in the steel industry may be cleaned and recycled, heavy metals like chromium, zinc, nickel and lead, sediments, volatile organic compounds (VOCs) and ammonia, amongst other pollutants, are produced and are usually not fully recovered for recycling. Contaminated sludge produced by the steel industry is still being sent to landfill, even in some European countries, countries with some of the best standards anywhere.

Mining, Oil and Gas

Mining and oil and gas extraction consumes relatively small volumes of freshwater overall (around 1% of the total in the United States), however the quantities consumed can be significant in the locations where mines and wells are located. Coal production typically requires 50 litres of water per tonne for open cast pits and 500 litres per tonne for underground mines.[28] More water is required where impurities must be washed from the extracted coal. Occasionally coal has been transported through pipelines in the form of slurry and this may require the abstraction of large volumes of ground water. Conventional oil and gas extraction uses very little water; however, the drilling of a typical hydraulic fracturing (fracking) well involves the injection of between 10 and 20 million litres of water into the hydrocarbon bearing rocks being drilled.[29] In order to exploit the 35,000 hydraulically fractured wells in the United States, it has been estimated that between 350 and 700 billion litres of water are used each year, enough to supply between 2 and 4 million people. In water stressed states, like Colorado, there are reports of increasing concerns expressed by local residents over the use of would-be municipal water supplies for fracking, with drilling companies outbidding farmers at state water supply auctions.[30]

When Will Water Consumption Peak?

This brief review indicates that farming, industry and city growth and city lifestyles have significantly increased demand for water. Per capita consumption grew rapidly after the Industrial Revolution, in concert with living standards, to reach a peak by the end of the twentieth century. However the population continues to grow globally, along with standards of living in less developed countries. These factors have, in turn, increased mining, food production and energy generation, all of which rely on water. The next chapter considers how water is supplied and the status of some of these sources.

4. Supply

The Roof of the World

Bam-i-Duniah (which means Roof of the World) is the term used in the Wakhi language (which is spoken in parts of Pakistan, Afghanistan, China and Tajikistan) to describe the region where the great mountain ranges of Central Asia meet. These mountains include the Himalayas, Hindu Kush, Kunlun Shan, Pamir and Tien Shan mountain ranges, which have been described as the 'water tower' of Asia. Glaciers and snowmelt from these mountains feed rivers, which are the main water supply for about 40% of the world's population. The Himalayas, which form the eastern section of these uplands, take their name from the Sanskrit for 'Abode of the Snow.' Snow falls and accumulates on these uplands during winter and melts in spring, feeding many rivers.[1] The Indus receives about 90% of its total discharge from glaciers and snow in its upper catchment and the role of snow is important for other major rivers including the Ganges, Brahmaputra, Yangtze and Huang He (Yellow River). Snowmelt also provides critical water supplies (between 75% and 85%) of the water used for agriculture and municipalities in Central Asia. Water from melting snow is also an important source of flows that drive hydroelectric power generation. As well as contributing toward river flows, snowmelt also infiltrates into soil and helps to recharge groundwater. Water supplies derived from glaciers and snowmelt, are especially vulnerable to climate change (see Chapter 5).

The Water Sensitive City, First Edition. Gary Grant.
© 2016 John Wiley & Sons, Ltd. Published 2016 by John Wiley & Sons, Ltd.

Figure 4.1 Satellite image of the Indus Basin. NASA.

Mountains

As well as places that trap snow, mountains are also places that tend to have higher rainfall than adjacent lowlands. As air is forced to rise and cool as it passes over mountain ranges, precipitation occurs. Importantly, this may occur in any season, including what may be a dry season for the adjacent lowlands. This rainfall, which feeds streams and recharges groundwater, may therefore be the main water source of water for relatively arid, downstream areas. An example is La Paz, in Bolivia, which has a relatively dry climate. There are 1.4 million people in that city, who rely on water flowing down from the adjacent Andean peaks. Mountain ecosystems, in particular forest soils, are also an important store of water. Organic matter in those deep forest soils tends to hold moisture. Water flows slowly into streams, ensuring year-round flows in most mountain areas. Water quality also benefits from its slow passage through forest soils, with sediments filtered and nutri-ents taken up by forest vegetation. The Ewaso Ng'iro River, which rises on the slopes of Mount Kenya, for example, receives 90% of its dry season flow from the mountain. This river was once perennial, however since the 1980s, there have been periods during the dry season when the river runs dry. With increasing population and water abstraction in the lowlands, it is now more important than ever that the natural ecosystems of Mount Kenya are conserved with, for example, illegal logging prevented, not only for the sake of its unique wildlife, but also to ensure that the water-supply problem is not exacerbated.[2]

Forests

The importance of maintaining forests in mountain areas for soil and water conservation purposes was recognized by the Swiss federal government in 1876, when it passed its first national forestry law. In the first half of the nineteenth century, a number of Swiss alpine forests had been clear cut to provide timber that was subsequently transported downstream, to be sold in France and the Netherlands. These activities exposed soils and this was believed to have contributed to a number of severe floods in the 1860s – floods that caused extensive damage to property and resulted in the loss of dozens of lives.[3] In the twentieth century, careful measurements of runoff in the Swiss Emmental showed that after a typical storm, peak flows were lower in rivers with forested catchments than in those in rivers with catchments without significant tree cover. Those classic studies, as well as others that have been undertaken since, have shown that forest soils tend to store and gradually release water, flattening out peak flows.[4] This characteristic is important for flood management but also to ensure that stream flows continue for longer in the dry periods between rainfall events.

Figure 4.2 The Queen Elizabeth II Reservoir, southwest London. Photograph by David Darrell-Lambert.

Reservoirs

The growth in population of cities and the expansion of agriculture has encouraged the creation of reservoirs, often in the upper catchment of river basins. Reservoirs are usually formed by building a dam across a river valley, although occasionally, quarries or other pits are flooded for this purpose. An example of a reservoir created by building a dam across of river is the East Branch Reservoir, built in 1891 on the Croton River, in order to store 20 million cubic metres of water for New York City.[5] Where the topography is unsuitable, particularly in low-lying areas, reservoirs are sometimes created through the construction of bunds or embankments, with water channelled or pumped into the artificial lake that is formed. An example includes the King George V reservoir, in London's Lea Valley, which was completed in 1913.[6] Another is the Queen Elizabeth II Reservoir in southwest London, part of the Walton Reservoirs complex, which was commissioned 1962. It is about 15 m deep, covers 128 hectares and holds 19,500 million litres.[7]

Impacts of Dams

Without the storage of large volumes of water in reservoirs of some kind, most modern cities would not be viable; however these installations have major impacts on river ecosystems. Rivers are important corridors for aquatic wildlife, most notably fish. Some fish, including many species of salmon and trout, migrate from the sea and swim up

Figure 4.3 San Clemente Dam, California. Removed November 2015 due to safety and environmental concerns. Photograph by Doug McKnight.

rivers to spawn in shallow, stony, upland reaches. Salmon are noted for their ability to swim through rapids and leap over barriers, however dams and reservoirs nearly always block their way. Where dams have blocked rivers, fish ladders and bypasses of various designs[8] may be used to allow fish to reach their spawning grounds, however these installations rarely allow the ease of passage previously provided by the natural river. Rivers carry sediments, usually in large quantities. These sediments are deposited in various locations along the full length of each natural river, with large volumes of sediment often accumulating in floodplains and river mouth deltas. Dams trap sediment behind them. As sediment builds up behind dams, it reduces the volume of water available and may eventually accumulate to a depth where the reservoir has effectively been extirpated. The concomitant reduction in sediment downstream can reduce the fertility of floodplains, leading to the loss of habitat and can also lead to coastal erosion. An example of this latter effect has been observed at the Godavari River delta in Andhra Pradesh, India, where a series of dams on the river has effectively diminished the delta at the river mouth and this, in turn, has led to the reduction in area of adjacent coastal mangroves, which are important for coastal protection and as nurseries for marine fish.[9] Where dams are high, reservoirs may reach a great depth, creating situations where stratification can lead to the creation of large volumes of cold water. Sometimes this cold water at the bottom of the reservoir can also become anoxic and turbid.[10] When this water is released, through the operation of hydroelectric plants or to maintain river flow, the low temperatures and low oxygen levels can have a damaging effect on the wildlife in the shallow downstream river waters. Occasionally water in reservoirs can become nutrient rich and when this water is discharged, this can have a damaging effect on aquatic life downstream, which may be adapted to water of a higher quality. The creation of reservoirs usually involves the flooding of forest, farmland and settlements upstream of the dam. The World Commission on Dams (a body formed by the World Bank and World Conservation Union (IUCN) in 1997 in response to growing opposition to large dam projects at that time) estimated that as many as 80 million people have been displaced by dams since the late 1950s.[11] In addition, important and irreplaceable natural resources, including primary forests, have been destroyed by dam building. It should be noted, however, that although dams are often built to create reservoirs of freshwater, the primary purpose of many of the larger dam projects is to generate hydroelectric power.

Lowland Rivers

Once rivers have flowed into the lowlands, water may still be abstracted, usually for irrigation but sometimes for municipal supplies. Wells or caissons are created in riverbeds where screens, sands and gravels are used to remove sediments from river water before it is pumped into service reservoirs. Many communities rely on abstraction from rivers and other surface waters including lowland lakes and ponds, however water quality is often poor, with contamination from sewers and agriculture a constant problem in most lowland locations. In such situations, levels of treatment are likely to be high and this means that the cost of drinking water abstracted from lowland rivers is relatively costly.

Licensing Abstraction

In most jurisdictions the abstraction of water from rivers is licensed according to legislation designed to protect river ecosystems and maintain water quality. The licensing of water abstraction is not only about the quantity of water that may be available, but also the effects that abstraction may have on water quality, habitats and species. In Europe, the Water Framework Directive has a major influence over the way that the authorities manage water abstraction.[12] The regulators or authorities usually require details of the abstraction method and equipment and a monitoring regime, which includes measurements of flow volumes. Normally the licensing regime will take account of all existing and proposed abstraction operations within a river basin as part of a comprehensive catchment management plan. Often permits or abstraction rights will pre-date the modern era and may not take account of ecosystem functionality. When licensing does take account of river ecology, a full description of the ecosystem and the way that it changes or has changed, may not be sufficiently detailed for the purposes of accurately predicting the impact of any increase in abstraction.

Aquifers

Where geological formations are suitable and water quality is good, groundwater may be abstracted. Where large quantities of water accumulate underground in permeable rocks, the formation is described as an aquifer. Some rocks may not have the permeability to allow water to flow freely and others may include minerals that dissolve, making the water too hard, too acidic, too alkaline or too salty. Aquifers are usually formed in chalk, limestone or sandstone formations. Groundwater flowing from such formations supplies springs and, depending on the changes in the underlying geology of a catchment, is often a major source of water that flows in watercourses. The abstraction of groundwater is often favoured over abstraction from rivers because water quality is usually good, with water having been filtered through rock formations, sometimes over very long periods of time, sometimes millennia. In general, deeper aquifers tend to have water that is of a higher quality because potential contaminants, which may enter the ground from the surface, are likely to be removed as the water takes a longer, slower route through the deeper rocks. Pollution at the surface can enter the soil and underlying rocks and eventually contaminate ground water. This can include point source pollutants, including leaking fuelling stations and landfill sites as well as pollutants from diffuse sources, including the fertilizers and pesticides routinely used on farms, often to excess.

Nitrate

Nitrate, in the form of nitrogen fertilizers, leaching into groundwater from farms, is a particular concern. In Europe, nitrate vulnerable zones (NVZs) have been delineated where groundwater is at risk from the leaching of nitrates into permeable rocks with important aquifers. Nitrate vulnerable zones were introduced under the Nitrates Directive in 1998. By 2009, NVZs covered about 70% of England. Within a NVZ, agricultural practices have to be modified in order to minimize nitrate leaching. An example is the restriction of the amount of animal manure that can be applied to grassland during the course of a year. The overall objective is to ensure that concentrations of nitrate in water do not exceed 50 mg/l, the maximum permitted under the European rules.[13] Groundwater represents about 35% of the mains water supplied by the water companies in England and Wales, so there is real concern over the risk of excessive amounts of nitrate entering groundwater.[14]

Overabstraction

About half of the drinking water in the United States is supplied from groundwater, along with the 250 billion litres of water used each day by agriculture.[15] It is becoming clear that the rate at which groundwater is being pumped from aquifers (not only in the United States but also in most other parts of the world) exceeds the rate at which aquifers are being replenished. This means that at some point in the relatively near future, depending on the rate of abstraction and the amount of rainfall received, wells will run dry. Over-abstraction also has an impact on the wider aquatic environment because soils can dry out, along with wet-lands and watercourses, which may rely on some groundwater for flows to be maintained. Another consequence of over-abstraction is that water quality in wells can be adversely affected. One common problem is saltwater intrusion. Most deep groundwater and groundwater under the coasts is saline and excessive pumping can cause saline groundwater to migrate inland or up into freshwater aquifers. This is a problem now being encountered in Cyprus, for example, where over-abstraction of coastal aquifers has led to saltwater intrusion. Water supply problems have dogged Cyprus for some time, with, for example, 8 million cubic metres of water being shipped from Greece in 2008, following a dry winter.[16]

Desalination

Cyprus has now embarked upon a desalination programme, with the intention of permanently solving its water supply problem. Desalination is the process whereby salt is removed from saline water (usually seawater but occasionally brackish estuarine water or ground-water). The process is attractive because seawater is extremely abundant (it makes up about 96.5% of all water on Earth) and the availability is not dependent in any way on precipitation. Desalination is an energy-intensive process, however, which makes it relatively costly. In 2013, there were around 20,000 plants supplying about 80 million cubic metres of drinking water per day to more than 300 million people. There are now desalination plants in more than 150 countries, with particularly large facilities in Saudi Arabia, the United Arab Emirates, Australia, California, Algeria, Israel and Spain.[17] The first industrial-scale desalination plants were developed during the twentieth century. The process was particularly important for the military, especially for ships and where large numbers of people needed to operate on remote islands, for example in the Pacific campaign of World War II. Many desalination plants still in use have old-fashioned designs, which are modelled on those early installations. These first generation desalination plants work by evaporating water from brine and distilling it. Saline water is heated (usually with steam) until it evaporates and then the vapour rises and condenses on a cool surface before being collected. Distillation is usually undertaken several times, in a multiple stage approach, in order to reduce heat losses.[18] Usually power generation is combined with distillation in order to further increase efficiency. Such power plants are often oil fired, meaning that the process has a large carbon footprint.

Reverse Osmosis

Reverse osmosis is a more recent development. In reverse osmosis plants, saline water is turned into freshwater by pumping the saline water through a semi-permeable membrane at high pressure.[19] Pores in the membrane allow the water molecules to pass through, but salt and other impurities are left behind. The process uses less energy than distillation (about 2.5 kWh per cubic metre of water compared with more than 100 kWh per cubic metre of water for distillation).[20] Virtually all new desalination plants use reverse osmosis and the majority of desalination plants worldwide are now of this type.

Impacts of Desalination

As well as being energy intensive, desalination has some negative environmental impacts. The most serious issue is the discharge of highly concentrated brine into the sea. This is believed to create hostile conditions for most marine organisms, although more research is needed to understand the full extent of changes to marine ecosystem that take place. As with any industrial process, there is also the possibility of pollutants being released, including substances that have been concentrated in the waste stream and chemicals and solvents used in the operation and periodic cleaning of the plant. There are also concerns regarding fish and plankton being trapped on plant-intake screens. This can be avoided, however, by creating intake points in chambers beneath the sea floor or beach and thereby allowing layers of gravel to prevent aquatic organisms from being trapped.

High Cost of Desalination

The relatively high cost of desalination means that this technique tends to be considered after other options have been ruled out; however, population growth and problems with disappearing aquifers, dwindling rivers and diminishing snowmelt mean that the amount of water supplied by desalination, both in total volume and as a proportion, is likely to grow rapidly in the coming years. The acceleration in its use is likely to be further increased if renewable technologies, like solar, wind and tidal power generation, can be coupled with desalination plants in order to reduce the carbon footprint of the process.

Rainwater Harvesting

In urban areas, where rainfall is adequate, rainwater harvesting can make a useful contribution to the supply of water, replacing some mains water that would otherwise be used for irrigation or toilet flushing and thereby reducing the amount of water that needs to be abstracted from the wider environment. Rainwater harvesting will continue to increase in sophistication and importance. More information on rainwater harvesting is included in Chapter 11.

Pressure and Pumps

Water supply depends on water distribution systems. Once water has been abstracted from the environment or has left a main reservoir, it is treated to the required standards to ensure that it is safe before being distributed across the city by a network of pipes. Depending on the topography, it may be necessary to create a series of elevated service reservoirs on local high ground or water towers in order to maintain the pressure and volumes required for each customer. In the United Kingdom, for example, water companies tend to supply water at 1 bar (equivalent to a 10m head), with a minimum standard of 0.7 bar set by the regulator, the Water Services Regulation Authority (Ofwat).[21] Reservoirs also help to provide additional water during periods of high demand and may be a useful temporary supply during pipeline maintenance. Water supply engineers work with the terrain wherever they can, however it may often be necessary to pump water to service reservoirs and water towers. Occasionally water supply companies need to pump water for considerable distances across districts or even between catchments to maintain supplies.

Pipework

Drinking quality water is distributed by pipework, ranging from 3.65m diameter pipes, which supply whole cities, to the 12mm diameter pipes that supply individual dwellings. Pipes must be durable to ensure a long life and watertight to prevent leakage and contamination from pathogens. Pipes can be manufactured from cast iron, steel, concrete, unplasticized polyvinylchloride (uPVC), polythene (including high density polythene or HDPE) and copper (and occasionally a few other less well known materials) and may be coupled using a range of methods including flanges, welds, compression joints or soldered sleeves. The World Health Organization provides detailed guidance on materials that are suitable for distributing drinking water.[22] Steel is comparatively expensive; however, it is strong and durable and can be easily connected (by welding and welded flanges). Although the relative amount of steel used in supply pipework is declining, most large diameter pipes are still constructed from steel. Cast iron is suited to use in pipes where water is at high pressure, however it is heavy and is supplied in relatively short lengths, which increases installation time and cost. Concrete pipes are corrosion free and durable. Concrete pipes are also relatively heavy, however, which makes these components more difficult to transport, handle and install. Polyvinylchloride pipes are noncorrosive, lightweight and strong. They are durable when below ground. When exposed to ultraviolet light they do eventually degrade and are more vulnerable to accidental damage than steel and concrete. Polythene, especially high-density polythene (HDPE), is an increasingly popular choice for water supply pipelines. HDPE is light-weight, durable, corrosion resistant, can bend and is easy to join. There can be problems with the expansion and contraction of plastic polymer pipes, however, especially when temperatures are above 65 °C. At the scale of the individual building, uPVC and copper pipes are most commonly used. uPVC is becoming increasing popular because of the ease of making joints using push-fit couplings, whereas copper pipes are traditionally joined with solder, which requires the use of a gas-fuelled soldering torch, an item of equipment, which becomes a fire hazard in careless hands. Another issue with copper pipes and fittings is that leaching of copper into water can occur. Although this is not a major public health risk, the United States Environmental Protection Agency has set an upper limit (a Maximum Contaminant Level Goal) for copper concentrations in drinking water of 1.3 mg per litre.[23] When water has been sitting in pipes for some time, it may be advisable to let the water run for a minute to ensure that any copper (or any other contaminants) that may have collected in the pipe are passed out before taking a drink.

Reliant on Rain

The supply of water is still largely reliant on precipitation, whether temporarily stored in mountain snow, being collected in reservoirs or abstracted from surface waters or ground water. All these sources are fed by precipitation. People in many regions of the world are therefore still relying on rain and are already vulnerable to the possibility of water shortages occurring because of drought. The next chapter considers how climate change may increase that risk.

5. Climate Change and Water

Climate Changes

The climate has always changed. For example the global warming which occurred around 12,000 years ago, brought longer, drier, summers to the Middle East. This favoured the cultivation of annual plants like cereals, which have grains, which can be stored for long periods. This breakthrough created the conditions for the establishment of permanent settlements, which subsequently grew in size. Since the end of the last ice age there has been an increase in global temperatures, although there have been many perturbations and reversals, including for example the Little Ice Age from the fourteenth to the nineteenth centuries.[1]

The Greenhouse Effect

Breakthroughs in understanding the role that the atmosphere plays in keeping trapping and moderating the heat of the sun began in the nineteenth century, when experimental practices and techniques became widespread and more refined. In the 1820s, the French physicist Joseph Fourier was the first person to explain the phenomenon now known as the 'greenhouse effect', whereby the atmosphere traps incoming solar radiation and keeps the earth warmer than it would otherwise be. In 1861, Irish physicist John Tyndall was able to show

The Water Sensitive City, First Edition. Gary Grant.
© 2016 John Wiley & Sons, Ltd. Published 2016 by John Wiley & Sons, Ltd.

how water vapour and carbon dioxide help to trap heat and in 1895 Swedish physicist Svante Arrhenius observed how the water and carbon dioxide molecules absorb infrared light.[2]

Callendar

In 1938, Guy Stewart Callendar, an English steam engineer and amateur meteorologist, showed that global temperatures had increased during the previous 50 years by analysing weather records, which people had started to gather in a systematic way across many locations during the nineteenth century. He was also able to link those increases to increases in the concentration of atmospheric carbon dioxide. Callendar's work was expanded upon by the Canadian, Gilbert Plass, in the 1950s. It is interesting to note that at this time, many people working in the field were fascinated by the phenomenon of the Ice Ages and were puzzled over the causes of sudden climate change, which had occurred in the relatively recent past. Most people at the time thought that global warming would be beneficial, helping to keep the much more extensive glaciers and ice caps that characterized the Ice Age, from returning. Callendar's contemporaries dismissed his work, however, questioning the accuracy and reliability of the data being used and doubting that concentrations of atmospheric carbon dioxide had increased over 50 years by the 10% that was being claimed.

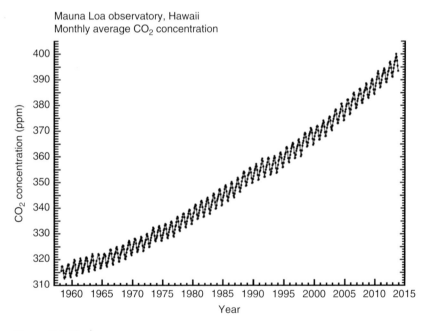

Figure 5.1 Keeling curve.

It would take meticulous observations over a sustained period to convince the majority of climatologists that carbon dioxide levels in the atmosphere really were increasing.[3]

Keeling

Those meticulous observations, required to bring about a tipping point in expert opinion began in 1958 on Mauna Loa, Hawaii (and to begin with also in Antarctica), when Dave Keeling of the Scripps Institution of Oceanography began measuring concentrations of atmospheric carbon dioxide. The Mauna Loa measurements continue to this day. Within a few years Keeling was able to show without any doubt that carbon dioxide levels were rising.[4]

Atmosphere and Oceans

Roger Revelle and Hans Suess undertook other seminal work at the Scripps Institution of Oceanography in the 1950s. Suess was able to show how it was possible to determine the amount of carbon dioxide in the atmosphere that was derived from burning fossil fuels by looking for the carbon-14 isotope, which is absent from coal and oil. There was a commonly held view at the time was that any surplus carbon dioxide produced by burning fossil fuels would be absorbed by the oceans in its entirety. Revelle was able to show that this belief was incorrect. Even so, most scientists still remained unconcerned about the possibility of rapid build up of greenhouses gases like carbon dioxide leading to future global warming. The prevailing view was that biological processes, including human activities, were far less important than geological processes, which operate in cycles that take millions of years. Civilization has been around for a few thousand years at most. How could such a relatively short-lived phenomenon (in geological terms) have affected the whole planet?

Details of the Carbon Cycle

During the 1970s researchers began to become more aware of other greenhouse gases in the atmosphere, notably methane, which traps much more incoming infrared energy and is therefore a more powerful greenhouse gas than carbon dioxide. Work was also undertaken on estimating and measuring the release of carbon dioxide through deforestation, agriculture and the production of cement. In addition, there was a blossoming of studies of the carbon cycle, as well as other geochemical cycles. By the early 1980s, developments in computer science and the availability of more processors and cheaper computer memory brought about a rapid growth in computer modelling. This meant that the modelling of the effect of increasing concentrations of greenhouse gases in the atmosphere became more sophisticated and much more rapid. There was also increasing agreement between the predictions of climate models and better correlation with observations of the climates of the past. Now able to see the results of many independent computer models, climatologists became increasingly alarmed about the possibility of rapid and catastrophic human-induced climate change.

The IPCC

This growing body of evidence on how carbon dioxide and other greenhouse gases was changing the atmosphere and oceans and the rising concerns about the potential for damaging climate change to affect civilization, led to the establishment of the International Panel on Climate Change (IPCC) in 1988. The United Nations Environment Programme (UNEP) and the World Meteorological Organization (WMO) came together to establish the IPCC in order to provide the world with a clear scientific view on climate change and its impacts.[5] In 1990 the IPCC produced its first report, which concluded that global temperatures had risen by between 0.3 and 0.6 °C during the previous century, that emissions by humanity are adding to the natural comple- ment of greenhouse gases and that further warming could be expected. At the 1992 Earth Summit in Rio, developing countries agreed to the target of returning their carbon dioxide emissions to 1990 levels. In 1997 the Kyoto Protocol was agreed. This new agreement required developed countries to reduce emissions by an average of 5% by 2012 (with a wide range of targets for individual countries). By this time, concerns were being voiced by industrialists and some politicians that reducing carbon emissions would damage the global economy. The United States Senate immediately declared that it would not ratify the agreement. By 2001 the George W. Bush administration had with- drawn the United States from the Kyoto process. In 2005, the Kyoto Protocol became international law for those countries still involved in the process. The next agreement, the 2009 Copenhagen Accord was a disappointing and controversial declaration agreed by the 192 coun- tries involved, with little progress on agreeing reductions in emissions.

Stern and the Financial Crisis

By 2009, the world was in the grip of a financial crisis and concerns about failing banks, debts, falling house prices and job losses meant that climate change was pushed down the news agenda. The Stern Review on the economics of climate change, published in 2006, was already being overshadowed by the financial crisis. The Stern Review was a landmark, 700-page report commissioned by the British Government and prepared in partnership with the Grantham Institute.[6] This document had warned that a great body of scientific evidence points to the likelihood of serious irreversible, runaway climate change if carbon dioxide emissions continue as part of a business as usual approach, with access to food and water and other natural resources likely to be badly affected and people in the poorest countries likely to suffer the most. Stern argued that the stabilization of carbon dioxide emissions was feasible as part of a transition to a low carbon economy and was compatible with economic growth. Stern also recommended the establishment of a global carbon pricing system, through trading, tax and regulation, in order to accelerate that transition to a low-carbon economy. Other advice for politicians and policy makers was that to stabilize atmospheric concentrations of carbon dioxide at between 500 and 550 parts per million would require the expenditure 1% of global GDP. It was argued that it was still possible to avoid the worst effects of climate change if strong, coordinated, collective, international action was to begin immediately. A warning from Stern was that delays would increase costs and increase the likelihood of crossing a threshold after which it will become too late to reverse the process.

400 ppm Breached

In 2014 the IPCC published its fifth assessment report.[7] By that time the mean concentration of carbon dioxide in the atmosphere had exceeded 400 parts per million and the rate of increase in emissions continued to be a concern to experts. In the period since the establishment of the IPCC in 1988, a wide range of scientists, working with data collected from earth-orbiting satellites and samples of ancient air trapped in ice cores from glaciers and ice caps, have been able to show that atmospheric carbon dioxide is now at its highest level for 650,000 years.

Two Degrees

Since the 1990s, there has been a common goal of limiting average global surface temperature increases over preindustrial averages to 2 °C, equivalent to 550 parts per million of atmospheric carbon dioxide. It has been thought that keeping temperature increases to 2 °C would mean that dangerous irreversible climate change would be

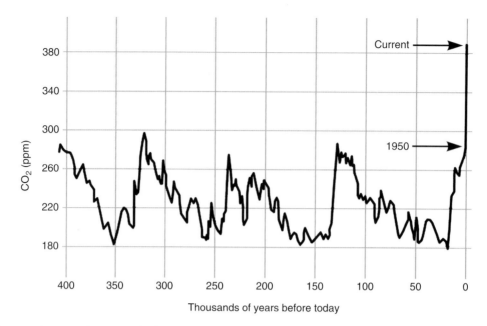

Figure 5.2 Atmospheric CO_2 for last 400,000 years. *Source*: NASA.[8]

avoided. In the years running up to the 2014 IPPC report, however, there were experts arguing that avoiding dangerous climate change is no longer possible, with the temperature increase already at 1°C and an increase to 550 parts per million likely to cause temperature increases in excess of 2°C. Given recent trajectories and delays in taking action, Kevin Anderson of the Tyndall Centre for Climate Change Research now predicts an increase in temperature of 4°C by 2060.[9] It seems that, although mitigation measures to stave off climate change are still urgently required, dangerous climate change is now almost inevitable and adaptation measures for these inevitable changes must be planned and implemented.

Sea Level Rises

One of those dangerous changes, which is already underway, is the rise in sea levels. Average global sea levels rose by 170mm over the course of the last century.[10] There are two mechanisms for the increase in levels. The first is thermal expansion of the waters and the second is the melting of land ice including glaciers and ice caps. The rate of increase has accelerated in the last decade and the predictions are for levels to continue to rise. The Third National Climate Assessment (NCA) produced by the United States government in 2014 predicts that sea levels will increase by between 300mm and 1200mm by 2100. A major difficulty for those predicting sea level rises is that the changes are unlikely to be linear, with an acceleration in rates of increase possible at some stage. Many researchers are predicting larger increases than the NCA's prediction in 2014, with an increase in sea level of 2 m by the year 2100 now considered more likely.[11] James Hansen, formerly head of NASA's Goddard Institute for Space Studies and now at Columbia University in New York, notes that paleoclimate records show that the last time the earth's temperature was 2 °C warmer than it is now (a threshold that we are expected to cross again during the coming century), sea levels were 6 to 8 m higher than they are at present. That would suggest that even the NCA's maximum 1200 mm rise in sea level by 2100 to be a huge underestimate of the amount of ice that is almost certainly yet to melt.[12] An indication of the difficulty for those modelling future sea level rises is the huge volume of water locked up in ice and uncertainties about how quickly this could melt. To provide a perspective on the volumes of water being measured and the scales involved in estimating sea level changes, if the entire Greenland ice sheet melted, for example, global sea levels would rise by 7.2 m.

Coastal Cities

Sea level rises are of particular concern for those hundreds of millions of people living in coastal cities. Major cities like London, New York and Shanghai will need improved sea defences if they are to survive future storm surges. It may be impossible to defend some cities from sea level rises, because of hydrological factors like high water tables, geological conditions that allow water to breach walls or estuarine or coastal locations which create a high exposure to storm surges. In addition there may be difficulties in raising money for defences or modifications from sceptical politicians and cash-strapped administrations before a catastrophe has occurred. Once a major catastrophe has occurred, there will be some cities where economic activity and social life has already been damaged beyond any reasonable chance of recovery. These possibilities suggest that there is a likelihood of large-scale migration as sea levels rise, with resettlement in new cities or enlarged existing cities somewhere on higher ground. New Orleans may be indicative of how many coastal cities may change. Eight years after Hurricane Katrina struck the city in 2005, the population of New Orleans had fallen by 24%, despite reconstruction efforts.[13] Those tenacious coastal communities that decide to stay and live with new watery conditions created by sea level rises will need to spend a large proportion of their income on adaptation measures. It has been suggested that Miami is the most vulnerable city in the world in terms of the potential for property damage caused by sea level rises, with more than $416 billion in buildings and infrastructure at risk. Difficulties for Miami include extensive low-lying areas and high water tables in the permeable limestone of coastal areas, where saltwater intrusion will also cause water supply problems.[14]

Warmer Seas

Much of the increased heat trapped by atmospheric carbon dioxide has warmed the oceans. The surface of the oceans absorbs most of this heat, with a significant increase observed in the uppermost 700 m of the water column.[15] The effects of warmer oceans include expansion (leading to sea level rises), an increase in the number of record high temperature weather events and an increasing number of intense rainfall events.[16] The increase in temperature in combined oceanic and atmospheric weather systems increases the water holding capacity of the atmosphere, which ultimately leads to heavier downpours in many places.

Ice

There is clear evidence that ice sheets have decreased in mass. For example NASA's Grace satellite has been able to show that the continent of Antarctica has been losing about 134 billion tonnes of ice per year since 2002, while the Greenland ice sheet has been losing 287 billion tonnes of ice per year.[17] Glaciers are also in retreat. There are photographs of many of the world's glaciers, in the Alps, in the Himalayas, Alaska and the Andes, taken a few decades apart in most cases, which show dramatic changes when compared with contemporary pictures. The Whitechuck Glacier in Washington, for example, has retreated by 2 km since the 1970s.[18] The extent and thickness of the Artic sea ice has also declined. Ice thickness over the central part of the Arctic Ocean has been reduced from an average of 3.59 m in 1975 to 1.25 m in 2012, a reduction of 65%. [19] The Artic sea ice reaches its annual minimum in September. Although the area of sea ice fluctuates, the overall trend is down, with a reduction from around 7 million km^2 in 1979 to around 5 million km^2 in 2014.[20]

Feedback Loops

The ice of glaciers and ice of the polar caps has a high albedo and reflects much of the sun's radiation. One of the problems with the decline in the surface area of ice is that more sea and land are exposed. Soil and sea have a lower albedo than ice and absorb more of the sun's energy, which leads to more warming. Positive feedback of this kind may contribute to acceleration of warming, an amplification which is usually difficult to model accurately. Other examples of feedback processes are associated with changes to soils, where warming can lead to drying, decomposition and wildfires, which can lead, in turn, to further release of carbon dioxide into the atmosphere. Another concern is that as the permafrost of the frozen north begins to thaw it may release methane as well as carbon dioxide. Methane is 20 times more potent as a greenhouse gas than carbon dioxide.[21] It has been estimated that 70,000 million tonnes of methane could be released from Western Siberia alone in the coming decades, contributing to a feedback process that could further accelerate warming. There is already evidence that the amount of methane being released is increasing.[22]

Ocean Chemistry

About 25% of the carbon dioxide released by human activity into the atmosphere subsequently dissolves into the oceans. This causes ocean acidification, a process that has been continuing since the Industrial Revolution. Since that time ocean acidity has fallen by 0.1 of the pH unit, which may not seem like much of a change, however pH is a logarithmic measure and a fall in 0.1 of a pH unit is equivalent to a 30% increase in acidity. Ocean acidification is predicted to have significant impacts on ocean life. Plants like seaweeds and sea grasses may benefit from higher carbon dioxide levels, however an increase in acidity in seawater will have a negative impact overall. There is particular concern for those species, which form skeletons or shells from calcium carbonate, including oysters, clams, sea urchins, corals and some forms of plankton. Higher acidity may dissolve their calcium carbonate skeletons and shells. Most of these groups of organisms are critical components of food chains, or, like corals, provide habitat for other species, often in ways that are not fully described or understood. The ensuing decline or loss of whole groups of organisms is likely to have a catastrophic impact on global fisheries, an industry that currently provides the protein for more than a billion people.[23]

Snowmelt

As described in Chapter 4, many cities in the world ultimately rely on snowmelt for their water supply. Mountain snow, known as snowpack, can effectively act as a reservoir of high-quality water, continuing to melt through the spring and summer and ensuring year-round flows in watercourses, ensuring that a reliable supply of water continues to reach centres of population during relatively dry periods. If mountain snow diminishes because of a decline in winter precipitation or if ice does not form because of an increase in autumn and winter temperatures, reservoirs of snow may be lost and rivers will tend to dry up in summer. In some locations there may be the option of increasing the consumption of groundwater, or on the coast desalination; however, the former may not be last and the latter is energy intensive. Snow in the Sierra Nevada mountain range is an important component of California's water supply system, providing a third of the water used by California's cities and farms each year.[24] In 2015, California was in the grip of a prolonged drought with April snowpack measurements the lowest since records began in 1930. Climate change is predicted to continue to reduce the volume and extent of the April snowpack in the future, diminishing water supplies for California's dry summers and autumns. Snowpack is important in California (as elsewhere where high mountains drain towards cities) because the state receives most of its precipitation during the winter months, when the snow settles, thereby forming a natural storage reservoir. In a typical year, California's snow-pack stores 18.5 billion cubic metres of water, more than all the water used in 2010 by California's cities. Snow accumulates at high elevations in the Sierra Nevada and southern Cascades from October to March and melts from April. Water from snowpack is important for those schemes that divert water from rivers upstream of the Sacramento-San Joaquin Delta as well as those that rely on the state's two major water supply networks, the Central Valley Project (CVP) and State Water Project (SWP). Together, the CVP and SWP supply more than 25 million people and more than 1.5 million hectares of farmland. Climate change is predicted to result in a decline in the snowpack of between by 25% and 40% by 2050, relative to recent historical averages. Climate change is also predicted to increase winter storms, which may fall as rain instead of snow, creating a higher risk of flood, with potentially less water available in the summer to meet demand. These changes are causing California's water supply experts to reconsider the existing outdated approach to water supply and storage. More, expensive, bigger dams of the kind already in use, which are being stridently demanded by some rattled by the current crisis will not bring a remedy. The existing infrastructure was designed for water storage in summer but will not provide the required flood protection during winter and spring. New surface water reservoirs are not the most cost effective, nor the most sophisticated way of addressing the changes being brought about by climate change. It seems that California should plan on the assumption that it will get less water from snowpack and will need to concentrate on smarter and more efficient water use and storm water capture in order to cope. The good news is that there is plenty of scope for efficiency improvements. For example more than half of the farms in California are still using inefficient, inflexible, gravity-based flood irrigation systems, which deliver too much water to crops and even draw water when it is not required.

Models and Projections

Climatologists and meteorologists are using computer modelling to predict how weather may change with climate change. These climate projections, as they are known, take years to produce and usually consider a range of scenarios based on assumptions of how much more carbon dioxide will be emitted. There are difficulties with projections, not only because we do not know how quickly and how effectively society's control of carbon dioxide emissions will proceed but also because there is natural variability in climate and the climate system is so complicated that even the most sophisticated models are imperfect representations. The Met Office Hadley Centre, United Kingdom Climate Impacts Programme and another 30 contributing organizations provide the United Kingdom Climate Projections. These projections of how climate will change during the twenty-first century are provided to inform planning by government and others. The 2009 climate change projections (UKCP09) are based on a model that breaks down the country into 25 km cells and makes probabilistic estimates for seven overlapping 30-year periods according to three future emission scenarios.[25] The United Kingdom Climate Projections estimate that by the 2080s (under a medium emissions scenario) all areas of the United Kingdom will be warmer (more so in the summer than in winter). There is a high level of uncertainty and variability in the predictions; however, most of the predicted changes will have a negative impact on society and the environment.[26] The increase in the mean summer temperature is likely to be in excess of 4 °C in southern England and 2.5 °C in the Scottish islands. The increase to the warmest day of summer, will vary considerably across the country depending on local conditions, but could be close to 5 °C. Changes in precipitation are likely to be highly variable with more rainfall in some locations and less in others; however, the prediction is for more precipitation in winter, with increases of up to 33% on the western side of the country. The far south of the country (the most densely populated and already the driest region) could see levels of precipitation fall by 40% in summer. Although summers may be drier overall, there is an increased probability of thunderstorms, which can bring very heavy rainfall for very short periods, which in turn will cause localized surface water flooding. Modelling of the future climate of England and Wales by researchers at the Met Office and the University of Newcastle indicates that short-duration summer rainfall events will become more intense and more frequent with climate change leading to more surface water flooding.[27] One of the difficulties for planners working on flood management is that summer storms can occur anywhere and can cause flooding in places outside of river flood plains or narrow valleys, in locations that have not been highlighted on flood risk maps.

Summer Storms

The chaos that summer storms can bring was brought into focus on 13 June 2015, when exceptionally heavy rains fell on Tbilisi, the capital city of Georgia. That particular event made the international news because many animals escaped from the zoo, including a tiger, which subsequently attacked and killed a member of the public. The zoo lost more than 300 animals, which were either drowned or subsequently shot. There were also reports of a penguin that was recaptured after swimming 60 km to the Azerbaijan border.[28] However what was not widely reported is that the flash flooding killed 19 people, injured 450 others (including 119 children) and caused damage to property and infrastructure estimated at $50 million.[29] Similar, often unreported, flash flooding occurs across the globe and will occur with ever more frequency as the atmosphere warms.

Heat Waves

Another summer or dry season phenomenon is the heat wave, a term usually used to describe a period of unusually hot weather, which may last a few days, but can stretch into weeks in extreme cases. The World Meteorological Organization defines a heat wave as 'a period of more than 5 consecutive days where the daily maximum temperature exceeds the average maximum temperature by 5 °C or more.'[30] High humidity heat waves are particularly unpleasant because they can create high night temperatures, which make it difficult for people to sleep and can exacerbate the symptoms of people who may already be infirm, particularly the elderly. Globally, heat waves take more lives each year than all the hurricanes, tornadoes, floods and earthquakes combined.[31] Climate change has already increased the number of heat waves that have occurred. For example it has been suggested that there is an 80% probability the record heat wave that struck Moscow in July 2010 would not have occurred without human-induced climate change. The Moscow heat wave saw temperatures exceed 40 °C (the average for that time of year is usually 23 °C); 5000 more lives were lost than would normally occur at that time of year, including 2000 who drowned while attempting to cool off in rivers and lakes.[32] Meteorologists are reporting that 10% of the global land mass is now experiencing extremely hot summers (compared to 0.1 to 0.2% for the period 1951–80).[33] And more heat waves are on their way.

Drought

Drought can threaten drinking water supplies and damage or even destroy whole ecosystems, including the watercourses and vegetation that help to keep cities cool. During droughts, food prices usually rise, which can also bring further hardships to the poorer residents of cities. Overall, drought has severe economic impacts, causing billions of dollars of losses every year in the United States.[34] In 2011, Texas had the driest year since 1895 and California had its driest ever year in 2013. In May 2015, drought continued to affect large parts of the western United States, Queensland Australia, Puerto Rico, southern and eastern Brazil, the southern Andes, the Sahel, Southern Africa and parts of India.[35] Meteorological drought occurs when rainfall is lower than is usually expected for a particular location, which makes general definitions difficult to agree.[36] Usually drought occurs when rainfall is unusually low or absent for months or years. Hydrological drought occurs when flows in streams, soil moisture, reservoirs or quantities of groundwater fall below particular levels. Climate change is predicted to increase evaporation rates and cause more precipitation to fall as rain rather than snow and in some places climate change will reduce overall precipitation, so the risk of drought will increase as global temperatures rise. There are regions where precipitation is already declining and this decline is predicted to continue with climate change. Affected areas include the Mediterranean, the southwestern parts of the United States, the Sahel and southern Africa. Cities in these and other affected zones will need to make a special effort to manage their water resources more carefully, however it seems likely that some failure to cope with changes is inevitable, with large-scale migration from affected areas likely to continue. For example, increasing drought in Africa's Sahel has been driving cross-border migrations, with agriculturists moving south within their own countries or crossing borders and often travelling further afield to join the unemployed or poorly paid crowding into the cities of Africa and Europe.[37] As well as planning for water scarcity, cities will increasingly need to prepare for the reception of increasingly large numbers of migrants, driven from their homelands by the increasingly common and increasingly severe droughts. Perhaps city planning will need to take more account of the projections of climatologists, to prepare for growth deliberately concentrated in those regions where rainfall is predicted to increase.

6. Microclimate

Climate

The climate is the average pattern of weather in a particular area over the long term, usually 30 years.[1] Averages and variations of temperature, precipitation, humidity and wind are recorded and analysed according to standard methods. Factors that determine climate include latitude, altitude, landforms (especially mountains) and proximity to water (especially oceans, which have their own circulation patterns, which may be subject to change).

Microclimate

Localized variations in climate also occur within recognized climatic zones. Microclimates, as the conditions in these smaller, distinct areas are known, occur when conditions are created that lead to local variations in precipitation, temperature, and humidity or wind speed. These areas may be many square kilometres in extent (for example a colder plateau or a fog-prone valley) or a few square metres (like for example a small park) or even a few square centimetres (for example in the shade of a tree or in the lee of a gatepost). The closer you look, the more microclimates you will find.

City Microclimates

Entire cities have distinct microclimates that may vary considerably from their hinterlands. In the United Kingdom, the duration of sunshine in cities is up to 15% less and there tends to be more cloud than in the wider countryside. Rainfall is up to 10% higher, yet humidity is up to 10% lower. Fog is twice as likely in the city.[2] Most cities in the world are hotter and drier than the countryside that surrounds them. Exceptions include a few desert cities where widespread irrigated planting can keep the urban area a little cooler than the surrounding areas. Cities also tend to be less windy than surrounding areas. Buildings act as a barrier to winds, although this varies considerably according to the spacing and size of buildings. Tall buildings can cause problems by creating eddies that create uncomfortable conditions for pedestrians at street level. Tall buildings can also throw long shadows and create rain shadows on lower roofs or streets.

Urban Heat-Island Effect

The most commonly cited microclimatic problem in cities is the urban heat-island (UHI) effect. This occurs when radiation from the sun is absorbed into the exposed concrete, masonry, brick and asphalt of the buildings and streets of the city. Some of this energy is reradiated from the urban fabric at night. Waste heat from vehicles, air conditioners and industrial processes may also contribute towards the problem. The lack of soil and moisture in cities means that evaporative cooling is reduced. The problem may be exacerbated by smog (a combination of air pollution and water droplets), which can form a dome, which traps pollutants but also helps to trap heat. Air quality and heat problems are more likely on calm days, when the trapped heat and smog may block cool breezes from entering the city from beyond the suburbs. Cities with more than a million inhabitants are typically a few degrees warmer than the surrounding countryside in both winter and summer and on hot summer nights some cities may be more than 10 °C warmer than adjacent rural areas.[3] The tall buildings and strong thermals caused by the urban heat island can also re-energize or intensify passing summer thunderstorms, thereby causing sudden intensified rainfall, although the effect may be experienced downwind and outside of the city limits.

Smog

Smog may be exacerbated by the urban heat island. Smog occurs when pollutants released by the combustion of fossil fuels combine with water vapour. The burning of coal caused the famous pea-souper smogs of London, which may have killed up to 12,000 people in the winter of

Hot

Cool

Black country boroughs are outlined.
Birmingham is to the east

Figure 6.1 West Midlands urban heat islands. Image provided by AECOM.

1952–53. Similar smogs are still occurring in cities in China.[4] The passing of the Clean Air Act of 1956 eventually ended the pea soup smogs in London, however the problem of photochemical smog still continues, in London as well as hundreds of other cities around the world. Places that are notorious for their recurring photochemical smogs include Delhi, Mexico City, Los Angeles, Tehran and Santiago (Chile). Cities that occur in valleys surrounded by mountains, including for example, Mexico City and Santiago, may suffer more because polluted air is less likely to be dispersed by breezes. Photochemical smog occurs when nitrogen oxides and volatile organic compounds (VOCs), produced by the combustion of fossil fuels, react in sunlight to form ozone and peroxylacetyl nitrate (PAN). Nitrogen oxides, VOCs, ozone and PAN are all toxic chemicals which occur together to form a brown haze that causes eye irritation, respiratory problems, heart and lung disease and cancers. Globally, air pollution in major cities was linked with 7 million premature deaths in 2012.[5]

Solving the Air-Pollution Problem

Solving the air-pollution problem will be primarily about reducing the production of pollutants, by replacing fossil fuel power generation and replacing petrol and diesel powered vehicles with new electric-powered or hydrogen-fuelled models; however, the problem of air pollution can be mitigated to some degree by increasing the surface area of vegetation in the city. Trees, shrubs and other plants have been shown to trap and absorb air pollution, including both particulates (soot) and gases. It is important to consider, however, that there is great variation in the ability of different species of plant to absorb air pollution. Research on this issue is still in its infancy, but the surface area of the plants, the density of leaf hairs and the stickiness of leaf surface are all thought to be important factors leading to the ability of plants to reduce air pollution.[6] Increasing the number of plants and the volume of soil and water in cities may not eradicate air quality problems, however all the indications are that this approach could rid cities of urban heat islands and solve most of the problems caused by harsh microclimate in particular locations. Infrared images of cities clearly show stark differences in surface temperatures. On a summer day when air temperature in the shade is 35 °C, it is common for the temperatures of asphalt roads and dark-coloured roofs to exceed 60 °C. Parks, gardens and green roofs tend to be at or below ambient air temperature. The deeper the soil and the more water there is, the cooler these areas become. Thermal images of green roofs reveal how even a few millimetres of extra depth of soil can reduce surface temperatures.

Cooler Roofs

Marco Schmidt of the Technical University in Berlin has looked in detail at the daily energy balance of conventional (bitumen covered) roofs and green roofs in summer in order to understand how these remarkable differences in surface temperature occur.[7] The first difference is in reflectivity (albedo). Although the difference is not as marked as it could be, green roofs reflect more of the sun's radiation than a conventional roof – about 60% more. Increased albedo is the reason that white roofs have been advocated as a solution to cooling cities. White roofs are conventional roofs, which have been covered with a white paint designed to reflect more of the sun's rays.[8] White roofs are an improvement over conventional roofs and are a wise choice when the weight of a green roof cannot be borne, however soil and vegetation offers much more. The soil and soil moisture of a green roof (or any green infrastructure element) allows evaporation to occur from the soil and evapotranspiration to occur from the plants and this provides cooling – the major factor accounting for the temperature difference between sealed and vegetated surfaces.

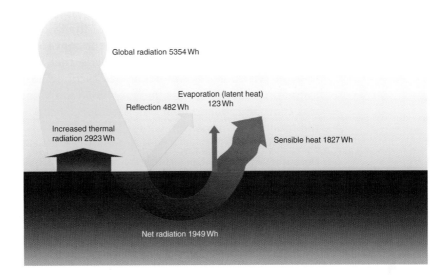

Radiation balance – black (conventional) roof

Global radiation 5354 Wh

Evaporation (latent heat)
123 Wh

Reflection 482 Wh

Increased thermal
radiation 2923 Wh

Sensible heat 1827 Wh

Net radiation 1949 Wh

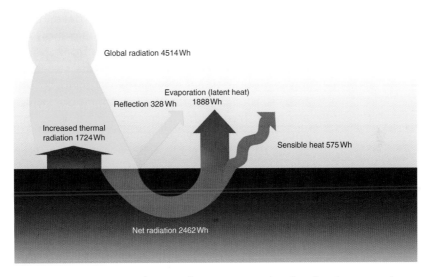

Radiation balance – green roof

Global radiation 4514 Wh

Evaporation (latent heat)
1888 Wh

Reflection 328 Wh

Increased thermal
radiation 1724 Wh

Sensible heat 575 Wh

Net radiation 2462 Wh

Figure 6.2 Comparison of energy flows on conventional roof and green roof - after Schmidt. Illustration by Marianna Magklara.

Living Walls

Living walls have also been shown to have a similar effect. Bernhard Scharf and his colleagues at the University of Natural Resources and Life Sciences in Vienna (known as BOKU) monitored a living wall on one of the city's buildings (a waste-management headquarters).[9] They found that, when compared with a conventional rendered wall, heat flux through the living wall was reduced by 50%. They also found that the surface temperature of the living wall in summer was 10 °C to 15 °C cooler than the adjacent conventional wall. The living walls also reduced temperatures inside that building by 5 °C. The cooling effect of the living wall was ascribed to the evaporation of irrigation water. What was unexpected in the Vienna living-wall study was that the reduced heat flux provided by the living wall also kept the building warmer in winter – unexpected because the high standard of insulation, which is typical of Austria, was thought to be sufficient to render additional insulation provided by vegetation ineffective and therefore unnecessary.

Trees Cool Streets

Below and beside the buildings there are the streets, and most pleasant streets have trees. When rainfall lands on a tree, a substantial amount of that rain is intercepted by the leaves of the tree – you may recall that dry spot on the pavement under the tree that you noticed when sheltering from a rain shower. The amount of rain intercepted by a tree canopy varies from species to species, but a certain amount of it never reaches the street, thereby also providing some additional evaporative cooling. It is the ability of street trees to provide cooler, shadier conditions, which is a better known attribute. Trees within a species vary considerably in size and spread and trees also vary considerably between species in terms the amount of light which is reflected or which filters through the canopy. The amount of transpiration and therefore cooling also differs, both between different species and also according to the availability of soil moisture. Most species of plants can vary their rates of transpiration by opening and closing stomata – the minute breathing pores on the surface of their leaves. In hotter and drier climates, it is thought that shade from trees is a more important factor in reducing temperatures than evaporative cooling. In California, for example, it has been calculated that shade provides two-and-a-half times more cooling than evaporation.[10] In more temperate climates, with higher rainfall, however, where there may be more soil moisture, which means that the role and shade in providing cooling from street trees is approximately equal in those places.[11] Detailed observations of urban trees in Tel Aviv showed that they intercept between 40% and 70% of the solar radiation that would otherwise strike the ground. In Tel Aviv, the temperature of streets shaded by trees at 3 p.m. (usually the hottest part of the day), varied according to the density of planting and the species, however some areas under *Ficus retusa* trees (the most effective species) were up to 3.4 °C cooler.[12]

Parks

Avenues of street trees can make a difference but groups of trees in parks and green spaces have a stronger effect. A park at midday at the height of summer may have surface temperatures which are between 15 and 20 °C cooler than adjacent built-up areas, giving rise to air temperatures between 2 and 8 °C cooler.[13] That characteristic is, in itself, beneficial for people seeking refuge in the park; however, the cooling effect extends out in to the surrounding area. For example, a large (500 ha) park in Mexico City (Chapultepec Park) was found to have a cooling effect 2 km from the edge of the park, that is a distance about as wide as the park itself.[14] In another study, a smaller park (60 ha in size) in Tama New Town, Tokyo was found to reduce midday air temperatures by up to 1.5 °C for up to 1 km from the park boundary.[15] Remarkably, the effect has been found for even smaller areas of green space. For example, the 0.5 hectare Benjamin Park in Haifa, which is only 100 m wide at its widest point, has an adjacent zone which is 1.5 °C cooler – a zone which extends for up to 150 m beyond the site boundary.[16]

Quality of Green Space

The quality of green space is also an important factor. Parks with too many paths or too much paving will not provide significant cooling. Very dry grassland may not be significantly cooler than adjacent urban areas. In contrast, groups of large trees with broad canopies will provide the most cooling. Large trees, especially native species, are also more likely to provide valuable wildlife habitats than small species chosen for their ornamental qualities. Like streets, many urban parks have been overprovided with unnecessary conventional drains, so that water that would otherwise have entered the soil to be subsequently taken up by the roots of trees, is sent to the surface water sewers and prematurely to rivers. Parks need sustainable drainage systems as well as the wider urban environment and parks can be the focus of new sustainable drainage systems that also benefit the wider urban environment (see Chapter 9). Bringing more water to the improved soils of urban parks in this way can support more trees, larger trees and better cooling for a neighbourhood.

Locating Trees

The locations of individual trees can have an important impact on the shading and cooling of an adjacent building. The walls of buildings shaded by trees have been reported to be 17 °C cooler than adjacent exposed walls.[17] In temperate climates too much foliage close to a building can mean the loss of valuable winter sunlight, so it is important to carefully consider the location of trees near buildings according to climate and latitude. In the higher latitudes, tall trees with lower branches removed may provide valuable midsummer shade of a roof whilst allowing winter sun to strike the façade. Also in higher latitudes, trees planted close to the west may provide useful afternoon summer shade without interfering with valuable morning and midday winter sun. Trees and shrubs (including hedges) planted too far away from a building to be able to cast shade may still be useful in creating favourable microclimate. Such plantings are known as 'climate-effect' trees (as opposed to the closer 'shade-effect' trees). Their main purpose is to act as a wind break and their location will be determined by the direction of the prevailing winds or winds that may be particularly cold and damaging to tender plants (northerly and easterly winds in many temperate parts of the northern hemisphere).

Water Bodies

Water bodies in cities have been found to be up to 6 °C lower in temperature than adjacent urban areas and can have a similar effect on microclimates as parks. The cooler water leads to lead to cooler air temperatures in the vicinity of the water body. Like parks, water bodies have a downwind cooling effect on summer temperatures in cities. It is commonplace for temperatures to be lowered by 1 or 2 °C in areas immediately downwind from water bodies during the day. However this differential is reduced at night with water bodies, which tend to lose heat more slowly than solid materials.[18]

Rivers

A study in Hiroshima, Japan, found that the cooling effect immediately above the Ota River reached 5 °C and that the zone of cooling extended nearly 100 m from the banks.[19] Studies of smaller rivers elsewhere have shown less cooling, especially when river water temperatures increase during the summer; however, rivers continue to be important even when water temperatures are high by providing an area with higher albedo. Rivers also promote evapotranspiration by supplying a ready supply of water to adjacent riparian and bankside vegetation. It is probably more useful to think as water bodies as part of the urban network of blue and green infrastructure, providing water for soil and vegetation as well as a range of other ecosystem services, rather than individual features with special performance characteristics in terms of cooling.

Heat-Related Deaths

Extreme temperatures have a serious impact on human health, so controlling microclimate is important. In temperate climates, cold is better understood than heat and although insulation is unsatisfactory even less is done to address the problem of excessive heat. In the relatively mild climate of the United Kingdom, heat-related stress accounts for an estimated 1100 premature deaths each year and more than 100,000 hospital patient days. When a heat wave hits a city, the situation can quickly escalate. For example the 2003 heat wave is estimated to have caused an additional 15,000 premature deaths in the United Kingdom and France. For each degree of temperature increase above the summer norm of 22 °C in London there are an extra 10 deaths in the capital.[20] The situation is even more serious nearer the equator. As I write this in May 2015, a heat wave in Andhra Pradesh and Telangara in India has sent temperatures above 47 °C, conditions that have already resulted in 1412 premature deaths.[21] People suffer and die during heat waves because of respiratory and cardiovascular conditions. Elderly people, pregnant women, people who are already suffering from respiratory and cardiovascular problems, the very young and people who are unable to leave their homes are most at risk. High night-time temperatures are a particular problem for these vulnerable people and high night-time temperatures are also associated with urban heat islands. Given that relatively small increases in temperature can result in the loss of life and given that climate change is predicted to increase the number of heat waves and to intensify them, there is a clear case for increasing the quantity and quality of green infrastructure in the urban core for the purpose of improving the microclimate. This will involve the interception, storage and reuse of rain water (and grey water perhaps) in order to allow vegetation on buildings and in streets to thrive but also to promote evaporative cooling. Parks and ground level interventions are important, however there also needs to be a stronger emphasis on vegetating roofs and walls in order to shield and cool the buildings where people live, study and work.

Energy Savings

Increasing the amount of shade and evaporative cooling by increasing the amount of soil and vegetation will also reduce the amount of air conditioning required, thereby saving energy and money. This effect has been clearly demonstrated in Chicago, where the thermal performances of City Hall (which has a green roof) and Cook County Hall (with a conventional roof) are continuously monitored. On hot summer days, the effect of the green roof is to slow down the warming of the building in the morning, to reduce peak temperatures and to accelerate cooling in the late afternoon. The maximum air temperature difference above the green roof as compared with the conventional roof is 9 °C. This cooling has had a significant effect on the interior temperature on the upper floors of City Hall and savings in air-conditioning costs are reported to be $5000 per annum.[22] The National Research Council in Canada has compared a 150 mm deep green roof with a conventional roof and found that the energy demand for cooling in the building was reduced by 75% with the green roof – equivalent to 6 kWh/m^2 per day.[23]

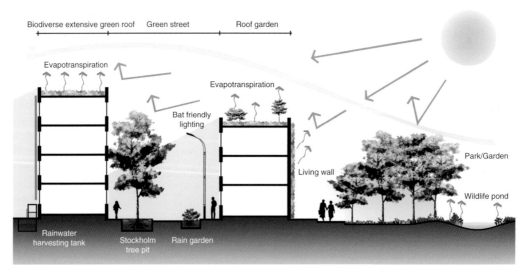

Figure 6.3 How green infrastructure improves urban microclimate. Illustration by Marianna Magklara. Based on an original by Luke Engleback.

An Overwhelming Case

The case for moderating the urban heat island and adapting to climate change by improving urban microclimates with blue and green infrastructure is overwhelmingly positive. A water-sensitive approach combines this effort with effective surface water management and the wise use of irrigation. Lives are extended, carbon dioxide production is reduced and money is saved. Air pollution is reduced and quality of life is improved.

7. Ecosystem Approach

The Great Acceleration

Civilization relies on the Earth's biological resources for its existence. Technology and the exploitation of fossil resources (fossil fuels and now fossil water) has 'turbocharged' our activities in a process that has been described as the 'Great Acceleration',[1] and has given many of us the impression that we have permanently risen above the constraints of everyday food webs. Yet we are still organisms – part of global biodiversity. Each of us is a single ecosystem, a microbiome, where the cells of bacteria and fungi outnumber our own cells by 10 to 1.[2] Beyond that we are part of ecosystems operating at various scales. Threats to species and ecosystems have never been as severe as they are now (certainly not since the human species emerged) and species extinctions caused by human activities continue at a worrying pace – what has been described as the sixth great mass extinction or the 'Anthropocene defaunation'.[3]

The Convention on Biological Diversity

A working group of the United Nations Environment Programme (UNEP) first considered the possibility of establishing an international convention on the conservation and sustainable use of biological diversity in the 1980s. Scientists and lawyers prepared the necessary documents, aware of the tensions that would emerge between representatives of

The Water Sensitive City, First Edition. Gary Grant.
© 2016 John Wiley & Sons, Ltd. Published 2016 by John Wiley & Sons, Ltd.

the developed and developing countries. Negotiations began and agreements to agree preceded the adoption of a text. The Convention on Biological Diversity (CBD) was opened for signature 1992 at the United Nations Conference on Environment and Development (the 'Earth Summit' in Rio).[4] By 1993 the Convention entered into force with 168 signatories – a major milestone in the transition from seeing the Earth as an asset to be exploited towards an understanding that we are part of a biosphere to be sustained. The CBD has three main objectives:

- the conservation of biological diversity;
- the sustainable use of the components of biological diversity;
- the fair and equitable sharing of the benefits arising out of the utilization of genetic resources.

The first session of the Conference of the Parties (COP), the governing body of the Convention was in 1994. The most recent session (COP12) was in 2014, where there were decisions relating to the Strategic Plan for Biodiversity for the period 2011–20, biodiversity and poverty eradication, gender considerations, engagement with business and local governments, tourism, invasive alien species, bush meat, human health, marine and coastal biodiversity, the emerging concern of synthetic biology and the issue of biodiversity and ecosystem services, as well as matters relating to finance, administration and efficiency.[5]

Ecosystem Approach

The primary framework for action under the CBD is the 'ecosystem approach', which was formally adopted at COP5 in Nairobi in the year 2000. In essence the ecosystem approach is a strategy for achieving the objectives of the CBD through integrated management of land, water and living resources. The idea is to promote conservation and sustainable use in an equitable way. The ecosystem approach is to be based on the application of scientific methods and a scientific understanding of ecology and the structure, processes, functions and interactions between organisms and their environment. There is recognition in the ecosystem approach that humans are an integral component of most ecosystems.

Ecosystems

Ecosystems are also defined as 'a dynamic complex of plant, animal and micro-organism communities and their non-living environment interacting as a functional unit.' This definition is broad and can refer to any ecosystem operating at scales from a particle of soil to the entire biosphere. There are difficulties and uncertainties because knowledge of the biosphere is incomplete and ecosystems are complex and dynamic. Ecological processes are often difficult to model and predict because outputs may not be directly proportional to inputs and there may be delays between the imposition of pressure and the observation of change (a phenomenon known as the lag effect), something which frequently causes surprise and fosters uncertainty in the minds of experts. What is observed may be counterintuitive and management must be flexible and adaptive in order to ensure that there are appropriate responses to experimentation, feedback and observation.

Principles of the Ecosystem Approach

Underpinning the ecosystem approach are 12 complementary and interlinked principles, which are paraphrased in the paragraphs that follow. Full versions can be seen on the CBD website.[6] A useful explanation of the principles, which are summarized below, has been provided by the United Kingdom's Joint Nature Conservation Committee.[7]

Principle 1 says that the objectives of management of land, water and living resources are a matter of choice for society. It urges that the rights and interests of local communities and indigenous peoples living off of the land should be recognized and that both cultural and biological diversity are central components of the ecosystem approach. It suggests that society's choices should be expressed as clearly as possible and that ecosystems should be managed for their intrinsic values and benefits for humans, in a fair and equitable way.

Principle 2 says that management should be decentralized to the lowest appropriate level. It is suggested that decentralized systems may lead to greater efficiency, effectiveness and equity. All stakeholders should be consulted. Where management is close to the ecosystem, the greater the sense of responsibility, ownership, accountability and participation with better use of local knowledge.

Principle 3 recognizes the interconnectedness of ecosystems and asks that ecosystem managers (meaning ecosystem managers in the widest possible sense, that is all of us to some degree) consider the effects (actual or potential) of their activities on adjacent and other ecosystems. Understanding the actual and potential effects of decisions may require new arrangements or means of organization for institutions involved in decision making. Compromises may be necessary, of course.

Principle 4 recognizes that there is an economic context for ecosystem management. It suggests that management programmes should reduce market distortions that adversely affect biodiversity.

It also recommends that incentives should promote biodiversity conservation and sustainable use. Principle 4 also asks that costs and benefits be internalized where feasible. In other words, the prices of goods and services should reflect the full cost to society and ecosystems. The ecosystem approach recognizes that markets often undervalue natural systems and provide perverse incentives and subsidies that favour ecosystem exploitation, that in turn causes losses to biodiversity and simplification of landscapes. It is noted that those who generate environmental costs (e.g. polluters) usually escape responsibility. Careful alignment of incentives, it is argued, will allow those who control the natural resource to benefit and will ensure that those who generate environmental costs will pay.

Principle 5 relates to the conservation of biodiversity through the conservation and restoration of ecosystem services. It notes that the preservation of species is not enough on its own and that ecosystem structure and functioning is essential, if ecosystem services are to be maintained and restored and species are to continue to have the habitats that they require. This principle seeks the recognition of the complex, dynamic relationships within species, among species and between species and their environment.

Principle 6 notes that ecosystems must be managed within functional limits and must not be exploited to destruction. Management objectives should take account of and avoid the creation of environmental conditions that limit natural productivity, ecosystem structure, functioning and diversity. Caution is required because ecosystem functioning may be affected to varying degrees by temporary, unpredictable or artificially maintained conditions.

Principle 7 recommends that the ecosystem approach be applied at appropriate spatial and temporal scales. Boundaries and timescales for management should be defined by users, managers, scientists and indigenous and local peoples and informed by an understanding of the interaction and integration of genes, species and ecosystems. Connectivity between areas should be promoted wherever possible.

Principle 8 recognizes that some ecosystem processes may be slower than others and that objectives for ecosystem management should be set for the long term. There may also be delays between actions and consequences – so called time lags. There is a conflict between the tendency for people to seek short-term gains over efforts to secure the environment in the long term and this needs to be managed.

Principle 9 recognizes that change is inevitable. It is understood that ecosystems change, including species composition and population size in any event. This means that management must be ready to adapt to change. The principle notes that ecosystems are characterized by complexities and uncertainties, which in turn lead to unpredictability. It also notes that traditional management of ecosystems may include disturbance, which may be important for beneficial ecosystem structure and functioning – management which may be counterintuitive but which may need to be maintained or restored. The ecosystem approach must include adaptive management in order to anticipate and adjust for changes and events. This means being cautious in making any decision that may preclude a full range of future options, but, also to consider mitigating actions to address long-term detrimental changes, such as the release of greenhouse gases into the atmosphere.

Principle 10 relates to the balance that must be struck between the exploitation and conservation of biological diversity. Biological diversity is critical for its intrinsic value and the key role it plays in providing the ecosystem services upon which we all ultimately depend. It notes that there has been a tendency in the past to manage components of biological diversity either as protected or nonprotected. There is a need to move away from this approach, where conservation and use are considered in a more sophisticated way. This implies a future where there is more of a continuum of management arrangements from strictly protected natural areas (which will continue) to human-made ecosystems (which can become more natural).

Principle 11 is about knowledge. The ecosystem approach involves the consideration of all forms of relevant information, including scientific and indigenous and local knowledge, innovation and practice. Better knowledge of ecosystem functioning and the impact of human activity is desirable. The principle urges all concerned parties and stakeholders to share and take into account all relevant information. Any assumptions that lay behind proposed management decisions should be made explicit and shaped by all available relevant knowledge as well as the views of stakeholders.

Principle 12, the final principle, is about inclusion. The ecosystem approach means that all relevant sectors of society and scientific disciplines should be involved. The management and conservation of biodiversity involves us all and there are many complicated, interconnected, interactions, effects and implications. Therefore this principle requires that all necessary experts and stakeholders at local, national, regional and international level be involved in the process.

Operational Guidance

In addition to the 12 principles, the ecosystem approach provides five points of operational guidance, which proffer advice for implementation. Again, as with the principles, the descriptions of the five guidance notes are paraphrased here.

The first point recommends a focus on the functional relationships and processes within ecosystems. The components of biodiversity, which are numerous beyond comprehension, control the stores and flows of energy, water and nutrients within ecosystems, and provide a resilience that allows some ecosystems and the biosphere as a whole to recover from major perturbations. Our knowledge and understanding of ecosystem functions and structure, and the roles of the components of biodiversity in ecosystems, is incomplete. More effort is required to understand ecosystem resilience and the effects of biodiversity loss (species and genetic levels) and habitat fragmentation as well as the underlying causes of biodiversity loss. There is also more to be understood regarding the processes and management decisions, which shape local biodiversity. The effect of organisms living together in ecosystems provides many goods and services of economic and social importance, which are required now and into the future. There may be a need to accelerate efforts to learn more about how biodiversity works; however, it is acknowledged that ecosystem management has to be carried out even in the absence of such knowledge. The ecosystem approach can facilitate practical management by ensuring that ecosystem managers (whether local communities or national policy makers) continue to learn as they continue their role as responsible stewards of the natural environment.

The second point relates to the sharing of benefits throughout humanity. Benefits that flow from biodiversity give us security and this needs to be sustained and, where appropriate, restored. The people responsible for the production of these benefits should receive their fair share. This will require improvements in the capacity of local communities to manage biodiversity and ecosystems. More needs to be done to bring about the proper valuation of ecosystem goods and services and the removal of perverse incentives that undervalue ecosystem goods and services. It is suggested that perverse incentives be replaced with mechanisms that encourage good practice.

The third point calls for adaptation. It is understood that ecosystem processes and functions are complex and variable and that uncertainty is increased by the interaction with human society and that these interactions need to be better understood. Ecosystem management must include a flexible approach to policy making and implementation, whereby management adjusts to take account of the lessons learned from monitoring. Adjusting to the unexpected is more useful than the pursuit of dogmatic programmes devised according to a misplaced belief in preconceptions. It is argued that ecosystem management needs to recognize the full diversity of social and cultural factors affecting the use of natural resource. Long-term, inflexible positions are likely to be inadequate or even destructive. Therefore ecosystem management should be a long-term operation that allows people to learn from experience. There is a strong need to improve scientific monitoring to strengthen the process.

The fourth point asks for management actions to be carried out at the most appropriate scale, with devolution of power to the lowest, most local level wherever feasible. Understanding the scale at which a particular ecosystem is functioning will help managers to define a problem or issue and formulate plans of action or interventions at the appropriate level or levels. Often, this approach will involve empowering local communities, which may have to assume new responsibilities, which may, in turn, require revised policy support and new legislation. People may also need to develop new capabilities. Where shared resources are involved, management decisions and actions may need to be organized in a way that engages a wide spectrum of stakeholders. This implies that new institutions may need to be created or existing institutions reformed in a way that facilitates decision making and, where necessary, conflict resolution. Although the focus should be on local communities, there are problems, which transcend catchments, bioregions and even national boundaries. People must therefore ensure that appropriate institutions and mechanisms are created to facilitate cooperation at all levels, from village to planet.

The fifth and final point continues with the theme of cooperation. It asks that we ensure intersectoral cooperation. Management of natural resources will require better communication and cooperation at every level of government and between agencies. It is suggested that special bodies and networks are created to ensure that this happens. The ecosystem approach should inform national biodiversity strategies and action plans. It is argued that there is a need to integrate the ecosystem approach into agriculture, fisheries, forestry and other activities that have an effect on biodiversity. I would add urban water to that list.

Ecosystem Approach and the Water-Sensitive City

How does the ecosystem approach relate to the future water-sensitive city? The first challenge is to make the link in the mind of the average citizen between the water that emanates from the tap or dribbles into the drain and water in the wider environment. It is the same water, but municipal or corporate control of water, which brings great convenience and comfort and improves health, creates a false impression that security of water supply is a remote, technical matter – the preserve of specialists and not part of democracy or the nature conservation agenda. The citizens of the future water-sensitive city will need to be more involved in the decision-making process that relates to water supply and drainage. In many jurisdictions, democratically elected local authorities should have more control over water management than they have at present, especially in places where water supply and treatment has become a privatized monopoly overseen by a centrally controlled and often remote regulatory body. Local authorities need to start cooperating more closely with river basin managers and water supply authorities. People can be more involved in the practical management of water, with individual households and businesses collecting and storing rainwater and having their own attenuation and cleaning facilities in the form of green roofs and rain gardens. Of course, despite everything that individuals can do, most of the hard major work of water infrastructure management will continue to be undertaken at a citywide or catchment level.

Impacts and Responsibilities

Most city dwellers probably need to be better informed and if necessary reminded of the impact of water abstraction and polluted discharges on the natural environment. Once people understand the link between their own water use and the condition of landscapes, rivers and oceans, they will begin to modify their behaviour. It will be unpopular, but charges for municipal water supply, drainage and sewage treatment should be increased to reflect the real costs. This will be another incentive for people to make some effort to become involved in the practical aspects of the urban water cycle. Prioritizing the preservation and restoration of ecosystem services will also change the way that designers and engineers deal with water. Any new scheme or initiative should be reviewed with the provision of multiple ecosystem services in mind. The conventional approach involves water being captured, purified, moved and disposed within a budget, with little or no consideration of biodiversity, ecosystems or amenity; however, nearly all water infrastructure installations and processes can do much more for nature and people and this will involve bringing more soil, vegetation and 'free' water into the city, in increasingly sophisticated and natural ways.

Limits

Whichever way the management of water in the future water-sensitive city is approached, it is important to recognize that there are limits to exploitation, beyond which ecosystems may be permanently damaged or destroyed. If water abstracted for a city results in a once-healthy river drying up altogether, or if effluent from a city creates a maritime dead zone, those cities then fall short of the standards expected by the ecosystem approach (specifically Principle 6 which asks that society recognizes and respects ecosystem limits). New sources of water or better waste treatment, which operate within the capacity of supporting ecosystems, are required in such cases. There will also be cities which are in unsuitable, unsustainable locations. Painful as it might be, it would be better to recognize that and relocate before catastrophe strikes and ecosystems are destroyed.

City-Scale Planning

The primary spatial scale for operating the water-sensitive city is the city itself. Many cities are well organized and have strong leadership (often more popular and more effective than national leadership) and this provides opportunities for citywide planning and action, where blue and green infrastructure, for example, can be organized into networks designed to provide more ecosystem services. There are also many opportunities for projects to be instigated and implemented in neighbourhoods, where champions within particular neighbourhoods can inspire others elsewhere, within the district, within the city but also in other distant cities. The city almost certainly buys water from some distance away and can become a more responsible consumer, working with others to protect and restore ecosystems within the catchment. In contrast to the typical short-term planning cycles of politicians and some businessmen, the planners of the water-sensitive city should be making long-term plans, looking at 50-year instead of 5-year programmes. It is possible to be too hasty; consideration should be given to the possibility of delays between the beginning of an action and the observation of change in an ecosystem – the lag effect. Change will be inevitable, of course, and monitoring should be put in place to enable the city to measure those changes and improve its ability to adapt.

The City Spectrum

Even within urban areas, there needs to be a spectrum of intensity of use. The blue-green infrastructure network will include places, which are more natural – often associated with watercourses – where the full range of ecosystem services and wildlife habitats can be provided. There will be places, like roads, for example, where the safe movement of vehicles and people will limit what can be done but even here there may be opportunities for greening. The balance between grey and green needs to be reset, with an expectation for more greenery being considered normal. Cities do not have to be as grey as they are now. There is much to learn about urban ecosystems, both the grey and green components. Our familiarity with cities has masked our ignorance of what the various components of the city are really doing. All assumptions should be described, understood, made explicit and challenged and all available knowledge should be shared. Creating the water-sensitive city will be a task for all.

Ecosystem Services

Principle 5 of the ecosystem approach urges us to prioritize the conservation of ecosystem services. Ecosystems provide various materials and processes that support life. These benefits, as they are described

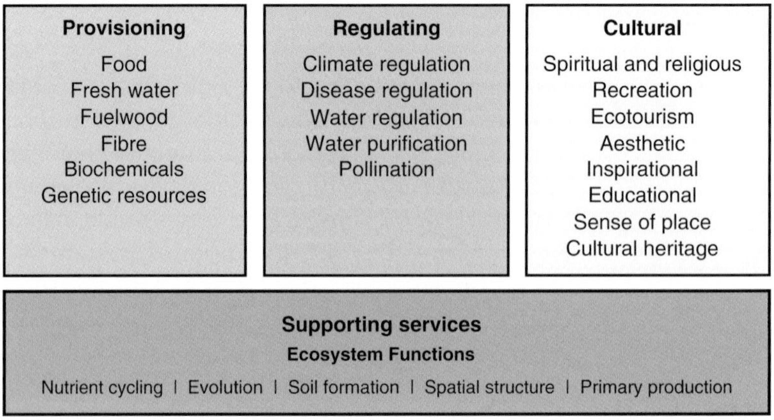

Provisioning	Regulating	Cultural
Food	Climate regulation	Spiritual and religious
Fresh water	Disease regulation	Recreation
Fuelwood	Water regulation	Ecotourism
Fibre	Water purification	Aesthetic
Biochemicals	Pollination	Inspirational
Genetic resources		Educational
		Sense of place
		Cultural heritage

Supporting services
Ecosystem Functions

Nutrient cycling | Evolution | Soil formation | Spatial structure | Primary production

Figure 7.1 Ecosystem services (after UN Millennium Ecosystem Assessment 2005).

when they support human life, are known as ecosystem services. Examples of ecosystem services include goods such as food and clean water and services such as the attenuation of rainwater, maintenance of climate and the conservation of soil. There are also nonmaterial benefits such as the provision of spaces for recreation and contemplation. Some ecosystem services relate to particular ecosystems and species. Examples of this include timber (e.g. mahogany from tropical American forests), foods (e.g. ocean fisheries) and medicines (e.g. penicillum mould). Other ecosystem services are a consequence of the operation of an ecosystem. These benefits are manifold, not fully understood and therefore more difficult to promote. A commonly cited example is that of soils, where largely unnoticed decomposition and nutrient recycling by soil microbes and invertebrates creates the fertility that supports plants and helps to clean water and air. Plants, in turn, support animals, of course, through the chain of herbivores and carnivores.

Valuation of Ecosystem Services

The interest in the economic valuation of ecosystem services has led some environmental economists to look at a relatively narrow stream of ecosystem services. They tend to focus on ecosystem services that are directly consumed or can be shown to support humans and human wellbeing. This is because it is difficult to measure and therefore value some of the ecosystem services that support or regulate in contrast to those goods (like food or timber) for which there is already a market, where prices can be validated and easily understood by policy makers and the general public. If a definition based on economic valuation is applied too strictly there is a risk that ecosystem service assessment could be biased toward services that are more easily quantifiable in monetary terms, with an incomplete and inadequate consideration of other services which may be critical for human wellbeing. Ecosystem services can be classified into supporting services, regulating services, provisioning services and cultural services, which are summarized in the paragraphs that follow.[8]

Supporting Services

Supporting services are the basic processes that underpin all others. All the other services depend in some way on these supporting services. Plants capture part of the electromagnetic energy arriving from the sun by the process known as photosynthesis. Most of the photosynthesis that supports cities occurs outside of those cities, but there is plenty of scope to increase the photosynthesis that occurs within cities. Nutrients and water are cycled through the biosphere. Again, most of that cycling occurs outside of cities, but plugging cities back into those cycles will improve those cities. Although these life-supporting processes have been described in essence, the way that they interact is poorly understood. The development of ecosystems, like soils for example, which provide supporting services, may take centuries and the cumulative effects of gradual degradation may take a long time to become obvious.

Regulating Services

Regulating services provided by ecosystems are difficult to define and may operate in a subtle way. They are also diverse. Examples include the way that pests and diseases may be regulated by natural processes, something that may be taken for granted whilst conditions remain relatively disease free but which has a profound influence on the supply of goods such as food, wood fuel and fibre. Climate is regulated and water is cleaned by ecosystems. Cities have tended to rely on artificial assistance for these services, particularly the cleaning of water, but nature can do much more for us than it does, even within the city's limits. As with supporting services, regulating services are interconnected and are associated with other categories of services.

Provisioning Services

Provisioning services are the most obvious. They include the products provided by ecosystems such as food, fibre, fuel (including peat, wood, vegetable oils and alcohol from plants) and fresh water. Many of these goods are provided by intensively managed ecosystems, such as farms and plantations, which may themselves be unsustainable, with a damaging effect on natural ecosystems. Other goods come from natural or seminatural ecosystems, for example ocean fisheries. Bringing farming or plantations into cities may be desirable in some respects; however, the ecosystem approach should result in the creation of more natural features that bring a wider range of ecosystem services than agricultural systems that concentrate on provisioning services. Urban farming is not enough on its own.

Cultural Services

Cultural services are provided from the landscapes within which people live and work. Nature is the setting; however, nature itself has been modified by centuries of human activity. Many cultural landscapes, like the rice paddies of China or the wood pastures of Europe, are highly modified but are biodiverse and thought to be sustainable. They are also associated with rich cultural heritages. Cities boast the various green and blue spaces such as parks and rivers which are places of recreation and relaxation but which also are the focus of cultural activities and the centres of many communities. These urban green spaces are now known to bring physical health, fitness and psychological comforts. Again, as is the case with provisioning services, spaces that provide only cultural services are not being worked hard enough to support us and to protect the wider environment. Under the ecosystem approach, supporting and regulating services should also be restored wherever feasible.

Economics and Ecosystems

Initiatives like The Economics of Ecosystems and Biodiversity (TEEB)[9] draw attention to the economic benefits of biodiversity and ecosystems and the costs of losses and degradation. The suggestion is that this approach will put a value on nature and will inform decision makers allowing them to factor nature into policies and accounting systems. This should have a positive educational effect; however knowing how volatile, distorted and arbitrary the pricing of ecosystem services can be, in the urban setting especially, nature will always tend be undervalued. In comparison with buildings and grey infrastructure, the monetary value of city green space is likely to continue to be insignificant. Single buildings can be worth hundreds of millions of dollars, whilst the trees in their shadows might be valued at a few thousand dollars at most.[10] The work of economists has its place; however, it is more important that protecting the biosphere is given the highest priority because our civilization will not persist without healthy ecosystems. As David

Suzuki has said, 'It makes no sense to elevate economics above the biosphere.'[11] Consideration of ecosystem services may be more useful in the context of evaluating the planning, design and operation of urban environments. We can ask these questions in order to test our proposals: 'Will this building or infrastructure element provide a range of ecosystem services apart from the cultural benefits, which may be its main focus? Will the project provide supporting services like photosynthesis or nutrient cycling? How will the project or scheme protect or restore regulating services like pollination, improve air quality, or reduce the effects of extreme weather? Can the project or scheme provide food, fibre, fuel or clean water? Will the project provide habitat for wildlife?' The citizens of future water-sensitive cities will be able to answer in the affirmative.

8. Rivers and Coasts

The Source

Until relatively recently in history, when mass manufacture made pipes and pumps affordable and trains and then, increasingly large goods vehicles, transformed transportation, cities were always established by a source of freshwater. A reliable, abundant source of freshwater is always required for drinking and washing and for watering livestock and irrigating crops. Rivers were important transportation routes, bringing traders, food and building materials. Riverside docks were the commercial hearts of cities, always crowded with boats, lined with warehouses, hotels and hostelries, suppliers of equipment and materials, with the merchants and bankers close by, perhaps a short walk away from the hubbub. The riverfront was often the main portal for the city.

A River of Life

Major rivers often support and connect several cities. An example is the River Danube, which connects Vienna (the capital of Austria), Bratislava (the capital of Slovakia), Budapest (the capital of Hungary) and Belgrade (the capital of Serbia). The Danube was the major transportation route for the Austro-Hungarian empire in the nineteenth century. In the year 1900, a project was conceived to connect Trieste, on the Adriatic coast, the most important seaport of the empire, with

The Water Sensitive City, First Edition. Gary Grant.
© 2016 John Wiley & Sons, Ltd. Published 2016 by John Wiley & Sons, Ltd.

the Danube, but it was never implemented. The Austro-Hungarian empire was broken up at the end of World War I but the Danube continues to play a key role in the life of the cities that line its banks.

Transport Revolution

During the nineteenth and twentieth centuries, the railway and road networks grew, and rivers and quays became less important. Where waterborne transport did continue, it did so with ever larger vessels, which led to major ports being moved downstream to estuarine and coastal locations with deeper berths. Quays become quieter and sometimes, derelict. In some cities, major new roads were driven along the waterfront, cutting off easy access to the river. Industrialization and population growth had also polluted many rivers to such a degree that they were no longer attractive.

Regeneration

An example of the decline in waterfront activity can be seen in New York City. Changes to the city's waterfront during the second half of the twentieth century can be seen by comparing aerial photographs of Manhattan and the East River and Hudson River, taken in 1951, when the shoreline bristled with piers and again in 1996 when most had gone.[1]

Once neglected to the point of dereliction, waterside and industrial properties, especially in major cities of the West, began to regain their allure when artists started to move in in the 1960s. Again, parts of

Figure 8.1 Aerial photographs of part of Manhattan in 1951 (left) and 1996. Images supplied by NYCityMap.

Manhattan, once believed to be dangerous and unloved were amongst the first places to see this colonization. High ceilings, plenty of light and low rents attracted bohemians to these city-centre buildings, which became dwellings, but also served as studios and meeting places. Typically, art galleries opened up on the ground floor of these former warehouses and factories and wealthy art collectors, charmed by what they found, started to buy and move in. These early waves of pioneers were subsequently followed by property developers, who would not only buy old industrial buildings to convert but would also build modern residential buildings in the industrial style in order to profit from the spiralling demand. Prices went up and whole districts were transformed in a process often described as gentrification, where places are created where only rich elites can afford to live, with low-income families and small businesses driven out.[2] The artists then move on to cheaper locations. This process continues wherever there is postindustrial charm and opportunity to be found, with planning authorities now attempting to duplicate these early well know successes in lesser known small towns and districts by initiating or supporting the establishment of galleries and special projects for artists and performers.

Water Quality and Regeneration

Water-quality improvements are a prerequisite for waterfront regeneration. The Allegheny River flows through the industrial city of Pittsburgh, Pennsylvania. The water was so badly polluted with sewage in the early twentieth century that there were typhoid outbreaks. Pollution from the steel industry then added to the problems. Despite this, even by 1923, local people were arguing that the riverfront should include public spaces to allow recreation. Industrial activities continued to dominate, however. Eventually, a plan was hatched to create some public parks. In the 1950s, 15 ha of industrial land was acquired at the confluence of the Allegheny and Monongahela rivers (which then become the Ohio River). This became Point State Park (known locally as the Point) in 1974. Miles of riverside trails and cycle paths followed and it has been claimed that this has engendered $4 billion of riverfront development, including a new technology park. The city is still working on the scheme, which was renamed as the Three River Parks plan in 2001.[3]

The Idea Spreads

Around the turn of the twenty-first century, officials in industrialized cities all over the developed world were taking a close look at their neglected waterfronts. The emphasis was usually on improving public access to rivers and quays by creating promenades and facilities for recreational use of the waterside. Such works were usually combined with important infrastructure improvements including repairs to river walls. There are usually historic buildings and structures that can be renovated to provide interest but the predominant mood has been for city governments to invest in large iconic modern buildings (usually cultural centres and galleries, for example on the River Clyde in Scotland[4]) and landmark bridges (like the Gateshead Millennium Bridge, in England).[5] Improvements to the public realm and transportation networks and the wide horizons and interesting vistas in turn have encouraged conversion of industrial buildings and zones to office and residential use. An example of this approach is the Hafencity in Hamburg, Germany.[6]

A More Natural Approach

Consideration of the restoration of natural features as part of urban riverfront restoration has been missing from the majority of these postindustrial schemes; however, with the rise of the concept of blue-green infrastructure and consideration of biodiversity in urban planning this is beginning to change, with more riverside parks, greening and softening of river walls and the restoration of riparian wetlands. An example of an ecologically informed riverside improvement is the 5.8ha site at Point Fraser on the Swan River in Perth, Western Australia, created between 2004 and 2006, where wetlands have been created to improve the quality of urban runoff and improve flood management but also to provide habitat for wildlife. Boardwalks make the site easy to access for all and there are places for people to sit and relax.[7] When the London Olympic Park was being planned in 2006, consideration was given to the replacement of the vertical walls of the River Lea and associated waterways with softer, vegetated banks. This was thought to be impractical and too expensive for what would become the busiest sections of the park between the main stadium and the aquatic centre; however, in the north of the park, the river was widened and gently sloping banks established in order to create a 'wetland bowl' with reed beds and adjacent seminatural terrestrial habitats.[8] In the more urban, southern part of the park, some wetland vegetation was established on shoulders at the foot of the vertical river walls so that, overall, the nature conservation value of the River Lea in this section was improved. As well as providing a naturalistic setting for the 2012 Olympic Games, these riverside features continue to form part of a permanent public park, the Queen Elizabeth Olympic Park, which is now the setting for large-scale development of commercial, residential and educational facilities.[9]

Figure 8.2 'Wetland bowl' on the River Lea in the Queen Elizabeth Olympic Park, London. Photograph by the author.

River Restoration and Urban Regeneration

In common with most riverside cities, Minneapolis in Minnesota has seen a decline in industry and river transport and closure of the railway yards that once covered several city blocks in the area adjacent to the River Mississippi. In 2010 the Minneapolis Park and Recreation Board began to promote riverfront development as a way of promoting cultural and economic development for the city-region. That sentiment in itself was not new; however, the initiative included a design competition, which placed a strong emphasis on the restoration of natural features within the river, the health of local people and the creation of a green economy. The winning submission by the TLS/KVA team outlined a comprehensive remediation of the city's storm-water management system, with a move towards stream restoration and the planting of biofiltration landscapes. The most radical element of the scheme was that riverside parks will work with natural processes, with the river being allowed to erode and deposit material along a more natural riverbank. The restored Northside Wetlands will include cycling trails and places for people to launch kayaks.[10]

Greening the River Wall

The traditional approach of creating vertical, armoured riverbanks makes sense where vessels need to be berthed, loaded and unloaded; however, in virtually all urban riverside locations piles or walls have replaced vegetated banks, even where access for boats is not required. This has resulted in loss of habitat for wetland species and the loss of spawning grounds for fish. The removal of wetland vegetation along riverbanks has also contributed towards the decline in water quality. Wetland vegetation traps sediments and through biological activity removes pollutants. Simply replacing hard engineered walls with gently sloping banks and vegetation is not always an easy task, however. Expensive urban development may have encroached into the historic floodplain, which limits the space available for the creation of a more naturalistic river edge and there may be insufficient space for the creation of a zone where the natural processes of erosion and deposition can be allowed to resume. It may also be unwise to create shallow slopes within the river channel because of the reduction in flow and storage capacity during floods. In these situations a compromise approach is to maintain or raise a flood-protection wall but to establish a vegetated zone between the top of the flood-protection wall and the water's edge. The vegetation within this zone can then be fairly natural in terms of species composition but the slope will be steeper than those vegetated slopes normally found in nature. This in turn requires the use of supporting structures to prevent erosion and collapse. Even with vertical features, it may be possible to use fenders or gabions to hold soil and vegetation in order to create a 'vertical beach', which can support a range of plants and animals which occur in the transitional zones between the high and low water marks, a technique that has been used at Deptford Creek in London.[11]

Coastal Cities

Most cities that are near the coast are not genuinely coastal cities in the sense that they have grown up on rivers in estuarine locations. Examples include Venice, which is located in a lagoon at the mouth of the River Po and River Piave and New Orleans in the Mississippi Delta. Coastal locations without rivers were unsuitable locations for major settlements during the period when people relied on watercourses for water supplies and needed quays for waterborne trade. Shallow beaches are unsuitable ports for large, laden vessels. The development of the railways and water-supply infrastructure in the nineteenth century meant that substantial settlements could be established in coastal locations that may previously have been only sparsely populated, usually by fishermen. These new conurbations grew up as resorts, places where working-class people could escape the poor air quality in the industrial cities where they lived, to begin with on day trips by rail and increasingly for week-long vacations, during periods when factories were shut down for annual maintenance. Blackpool in Lancashire, England, grew up as a resort, serving people in the industrial cities of that county. North Sea resorts, including Bridlington, Cleethorpes and Skegness amongst others, provided a similar service for people in the industrial cities in Yorkshire. A similar pattern of resort development was established in other northern European countries, with German workers heading for a 'Sea Spa' (Seebad) on the Baltic or North Sea coasts. One example is Binz, the largest resort on the island of Rügen, which was popular with Berliners. Similar resorts grew up in Belgium, the Netherlands, France and Italy, servicing people coming from the industrial cities of those countries. In the United States, beach resorts grew up close to the major cities as rail links improved. A particularly famous example is Coney Island, a barrier island on the Atlantic coast near New York City, which flourished in the early twentieth century.[12]

Beach Life

These nineteenth- and twentieth-century resorts nearly all have beaches providing relatively easy access to the sea for bathers, with nearby accommodation (hotels and guesthouses) and entertainment, typically theatres, amusement parks and places to eat and drink. One of the features of many of the nineteenth-century resorts was the pier, which provided more space for people to walk (promenade), space for places of entertainment and occasionally a useful berth for passenger boats.[13] During the last decades of the twentieth century, as air travel grew in popularity and people began to travel to southern Europe for their holidays, the coastal resorts of northern Europe fell into decline. These towns became popular places for people to retire to, to live modestly on pensions. This has added pressure to healthcare providers and social services. The abundance of inexpensive accommodation in some resorts has attracted the unemployed and others who rely on state support.[14] Small businesses have failed as footfall from tourists has dropped and disposable income has fallen in the local population, which has all contributed to a spiral of decline.

Fun in the Sun

As the northern seaside resorts declined, those in the sunnier climes of southern Europe grew. Starting in the 1960s, the Spanish Mediterranean coast became a magnet for holidaymakers from Britain, Germany, Scandinavia and other northern European industrialized countries, where wages were growing rapidly. Villages became cities in a few years and an almost continuous coastal strip of development became established around the Spanish coast. Torremolinos on the Costa del Sol, in southern Spain, for example, one of the first resorts to be established in that region, had a population of a few hundred people in the nineteenth century. This increased rapidly during the second half of the twentieth century to reach its current level of 70,000. Development was chaotic and poorly planned, with a preponderance of high-rise hotels and the loss of amenity. This pattern was repeated in coastal towns around the Mediterranean. Deterioration in the quality of the built environment in these resorts, which was often accompanied by damage to the beach itself, encouraged people to seek new destinations. The demand for new, quieter and cleaner destinations has encouraged villages and regions throughout the Mediterranean to enter this market, with airports and resort towns springing up in Greece, Cyprus and Turkey and more recently Morocco and Tunisia. As intercontinental travel has become more affordable, people are now travelling even further afield. Resorts in Thailand, for example, are attracting up to several million tourists each year, with visitors from China, Europe, Russia, the United States and Australia. Many coastal towns have attempted to reverse the decline in fortunes, with new attractions, including art galleries, centres of gastronomic excellence, historic buildings, water sports and adjacent areas of scenic and natural beauty, amongst other schemes; however, overall the long-term trend in the older resorts, especially those with poor connections to the largest cities, is for further decline.

The Front Line

Wherever they are, however, coastal locations, especially beaches but also cliffs, are naturally unstable. Shifting sands create problems, threatening buildings and other structures, which are close to the shore and when large numbers of people gather, moving around on a soft ground above the high water mark, further erosion can occur. This has meant that, as resorts became popular, boardwalks and masonry (and later concrete) promenades were built. Promenades are nearly also associated with an adjacent, parallel, road. The strengthening of the shoreline often accelerates the erosion of the beach by longshore drift, which leads to the installation of rock, timber or concrete groynes, in front of and perpendicular to the promenade, to prevent the loss of sand.[15] The presence of hotels and other buildings close to the beach and the routing of some of the most important streets in a typical seaside town behind the promenade mean that there are limited options for the town to retreat from the sea, in a way that would provide space for softer sea defences, like a wider, higher beach. Planners and engineers are left with more expensive, heavily engineered options. It is not only low-lying coastal areas that have problems. Where coastal towns are located on or near cliffs or steep slopes near the shore, with softer rocks, in particular, there can be problems with landslides and subsidence. A combination of heavier than usual rainfall, combined with accelerated coastal erosion, can result in the destruction of buildings, roads and other infrastructure.

An Uncertain Future

Coastal cities, then, usually have the twin problems of the changing tastes of holidaymakers combined with the constant wearing away of the shore by wind and waves. The latter problem will be exacerbated by climate change, which will cause sea levels to rise. These problems will be compounded by changing rainfall patterns, which can lead to increases in waterlogging and instability of slopes. As the economic performance of these places fades, less money will be available to repair crumbling coastal defences and failing water-supply infrastructure. Although the costs of infrastructure are usually met by regional or national governments, the large number of vulnerable coastal cities in some jurisdictions means that difficult choices will need to be made as budgets are squeezed at a time when the rates of erosion accelerate. In the coming decades, many coastal cities, especially those with declining fortunes in places where storms are more frequent and more ferocious, will be abandoned.

9. Near-Natural Drainage

Rain-Garden Origins

In 1990, Dick Brinker, a housing developer building Somerset, an 80 acre (32 ha) site for 199 new homes in Prince George's County, Maryland, in the United States, had the idea of replacing four ponds with a series of landscape features, which he felt would be more appropriate in the circumstances.[1] The US Clean Water Act (CWA) in 1977 had referred to Best Management Practices (BMPs) to control water pollution and, in urban situations, the response to this tended to be the creation of sizeable ponds fed by conventional drainage. An overreliance on ponds by some engineers is still a problem now, indicating how long it takes for new ideas to spread through the professions. Back in 1990, however, Brinker was aware of examples of industrial and commercial sites where space was limited, where landscaped areas had been modified to allow water to pool for a short time before infiltrating into the soil. Brinker and his daughter Theresa approached Larry Coffman, who was an official at Prince George's County (the local drainage authority), to seek to modify the drainage plans for the Somerset site, replacing kerbs, pipes and ponds with shallow basins filled with free draining soils in each building plot. Hanifin Associates, consultants to Prince George's County, dubbed these infiltration features 'Rain Gardens'. The rain gardens were only 25% of the cost of a conventional drainage and ponds and the extra space that was released meant that the developer was able to add 'six or seven lots' to the scheme. The project was monitored for 2 years, which demonstrated its effectiveness; in 1993 the county published a

The Water Sensitive City, First Edition. Gary Grant.
© 2016 John Wiley & Sons, Ltd. Published 2016 by John Wiley & Sons, Ltd.

design manual for bioretention in stormwater management. The most recent version, published in 2007, entitled 'Bioretention Manual', is still available.[2] In 1998, Larry Coffman worked with others to start the Low Impact Development Center to promote the concept of Low Impact Development (LID).[3]

Scotland Takes Up the Challenge

When, in the late 1990s, the town of Dunfermline in Scotland was looking to expand to the east with the construction of 5000 houses and associated infrastructure, the high cost of conventional surface-water drainage infrastructure became an obstacle. Brian D'Arcy, an expert in diffuse aquatic pollution working at the Scottish Environmental Protection Agency (SEPA) at the time, proposed an approach that would mimic natural drainage. In 1997 a sustainable urban drainage system (SUDS) working party was established and by 2000 CIRIA (the Construction Industry Research and Information Association) had published a SUDS manual for use in Scotland and Northern Ireland. The Scottish Government issued an advice note on planning and SUDS in 2001. The LID approach had crossed the Atlantic. By 2006, the Scottish Government used its obligations under the European Water Framework Directive to improve water quality to require SUDS for new development under General Binding Rule 10.[4]

England & Wales

Progress in England was slower. Sustainable urban drainage systems became sustainable drainage systems (SuDS) to acknowledge that the approach could be applied anywhere. The Flood and Water Management Act 2010 brought the concept of SuDS into the legislation;[5] however, concerns over increased costs and responsibilities for maintenance meant that implementation of the Act was still being delayed in 2014. Nevertheless, by that time, supported by guidance provided by CIRIA (e.g. the SuDs Manual 2007),[6] planners, engineers and designers in England and Wales were aware of SuDS and most new developments were incorporating some elements of SuDS, even if not every scheme could be said to demonstrate a full commitment to the approach.

Working with Nature

It is possible to characterize conventional drainage, with its pipes and sealed surfaces, as fast and dirty drainage and the more natural approach envisaged by the proponents of low impact development, sustainable drainage systems and 'near-natural' drainage in Germany[7] as slow (and clean) drainage, but what do these near-natural drainage concepts involve? The first overarching philosophy is that of the ecosystem approach (see Chapter 7). In brief, the ecosystem approach is for people to work together in a holistic way, working with nature and applying ecological knowledge to solve practical problems. It certainly helps if rainwater is thought of as an asset rather than a liability. The main objective of low-impact development is to reduce runoff rates and maintain water quality at predevelopment (greenfield) levels. This is achieved by seeking ways of intercepting and slowing rainfall as it makes its journey from rooftops to watercourses, with a preference for means of intercepting rainfall at source. The idea is to store runoff and release it slowly (a process known as attenuation); to allow water to soak into the ground (where ground conditions are suitable) – a process known as infiltration and to move water on the surface as required (conveyance). Pollutants can be filtered out as part of the process and sediments can settle out in purpose-made features before they reach watercourses. It is also important that water quality, amenity and biodiversity are all fully considered as well as the reduction in the volume of water leaving a site.

Management Train

Advocates of sustainable drainage systems describe a management train (or treatment train), where there is consideration of the movement of rainfall and surface water through a catchment. The process starts with prevention, where efforts are made to reduce sources of water pollution and the overall area of impermeable surfaces reduced to a minimum. Look at areas of impermeable surfaces, including, for example roads, pavements and roofs. Can these areas be reduced in size or covered with soil and vegetation? Roofs can often be greened. Paved surfaces can usually be reduced in area. Sources of pollution, like vehicles, can be modified so that they produce less pollution. Activities, like the washing of vehicles, can be prohibited, modified or moved to places where runoff is intercepted and cleaned. Canopies can be created over places where pollutants are produced and dedicated drainage and treatment facilities

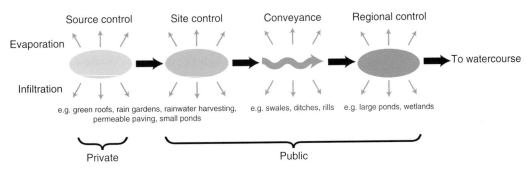

Figure 9.1 The SuDS management train. Illustration by Marianna Magklara.

created for activities that generate or use significant volumes of pollutants, including for example, litter, oils, detergents, fertilizers, pesticides and herbicides. Surface water drainage systems designed to manage storm water should never receive untreated sewage or grey water (water from baths, sinks or washing machines).

Source Control

It is best to capture and slow rainfall where it falls, locally, an approach known as source control. Examples of source control include green roofs, permeable surfaces and rainwater harvesting (rainwater harvesting is described in Chapter 11). Good source control involves many relatively small and inexpensive interventions, which can reduce the size of components required further downstream. Source control, however, is often installed on private property and requires investment and maintenance by the private owner, who might be reluctant to make such a contribution, even if it saves money overall and is for the greater good.

Green Roofs

Green roofs must be a priority for anyone creating a near-natural drainage scheme. Green roofs include intensive green roofs (also known as roof gardens) and the relatively lightweight, low-maintenance, extensive green roofs. Less common are blue roofs, which are designed to store rainwater for slow release (attenuation), evaporative cooling or for later use in irrigation. Water in blue roofs can be open to the elements, stored in permeable media or held beneath a deck. Green roofs usually consist of a free-draining yet water-absorbent growing medium (substrate). Substrate usually includes porous stony material like pumice or crushed brick. This combined with modest amounts of organic material means that the substrate remains free draining yet can hold water. Free draining materials that cannot hold water, like sand or crushed concrete, are not normally used on green roofs, for they would need to be combined with a high proportion of other water-absorbent materials. In order to reduce the risk of fire to insignificant levels, the proportion of organic material used on extensive green roofs is restricted to 20% or less under the German standards[8] and this has been copied in many other countries, including the United Kingdom.[9] The main objective is to keep plants alive in what can be stressful situations; however, green roofs can also be fine tuned in order to maximize the volume of rainwater retained on the roof – an objective that is, of course, compatible with that of

Figure 9.2 Biodiverse, wet, extensive green roof on St. Gallen Hospital, Switzerland. Photograph by the author.

providing vegetation. Another complementary objective is to provide water, which will evaporate or transpire through plants in order to help keep the building cool in summer.

Holding Water on the Roof

A simple approach to maximizing the water-holding capacity of green roofs is to increase the depth of substrate. On heavy structures like steel-reinforced concrete this is certainly an option, although in the case of most buildings with relatively lightweight roof decks, there will be a limit to the weight of green roof that can be carried and the concomitant quantity of water that can be stored in the substrate. The amount of rainfall that can be stored by green roofs varies according to the level of moisture in the substrate, which varies considerably from season to season. In winter, a series of rainfall events can saturate a typical green roof. In summer it is more likely that a green roof will be drier and therefore capable of intercepting more rainfall. The role of green roofs in helping to reduce the potential for summer storms to cause localized flooding is therefore very important. Several institutes have now measured the ability of various green roofs to intercept rainfall. What has been shown is that extensive green roofs with 100 mm of substrate intercept the majority (70% to 80%) of rainfall events.[10] Typically with rainfall events below 20 mm, no rainfall leaves the roof; however for 10% of the rain days in each year, when roofs are saturated or when rainfall is particularly heavy, roofs begin to shed most of the rain that falls upon them.[11] The voids in green roof substrates, which are usually between 15% and 30%, mean that there is space to store considerable volumes of water in the deeper soils of intensive green roofs. In the shallower substrates of extensive green roofs, typically 50% of the water that lands of the roof will be stored and subsequently lost through evapotranspiration over the course of a year; however, in the deeper soils of roof gardens, this can be nearer to 90%.[12] Clearly there is a case for carefully designing green roofs for the attenuation of rainfall; however, there will nearly always be times where water leaves the roof.

Rain Gardens

Once water has left the roof, there will normally be further opportunities to pass the water through soil before it needs to be carried away from the building or its curtilage. Following the seminal work undertaken in Maryland in the 1990s, as described at the beginning of this chapter, the idea of the rain garden spread to many other cities in the United States. For example in the City of Portland, Oregon, where runoff from urban development threatens the quality of water in the River Willamette, which supports a population of salmon, which are notoriously intolerant of pollution, widespread use of rain gardens was promoted.[13] Rain gardens can improve the quality of urban runoff; however, they can also reduce the volumes of water entering combined sewers, which when overwhelmed by storm water can overflow and carry sewage into the river. The City of Portland has performed flow tests on some of its rain gardens. In 2003, the Glencoe Rain Garden was subjected to flows that simulated a 1 in 25 year short-duration, high-intensity summer storm. This meant that over a period of 10 hours, 15,600 US gallons (about 59 m^3) of water was poured into a rain garden, which covered 5% of the subcatchment being serviced. Peak flows were reduced by 80% and delayed by 28 minutes.[14] What this means in practice is that for most rainfall events, no water leaves the rain garden. Rain gardens are usually between 10% and 30% of the area that they are designed to drain. They are shallow basins normally no more than 150 mm deep, filled with free-draining yet water-absorbent soil (usually a blend of

Figure 9.3 Rain garden in Portland, Oregon. Photograph by Dusty Gedge.

coarse sand and organic matter). There is usually a freeboard that can allow the basin to flood for one or two hours following particularly heavy downpours; however, rain gardens are usually dry and plants used need to be tolerant of temporary waterlogging and long dry spells (there is a surprisingly long list of plants in this category in most places). Rain gardens are not normally wetlands. Once full, rain gardens may drain into swales, conventional drains, or other sustainable drainage features. The authorities in the United States usually recommend that rain gardens are kept at least 10 feet (3 m) away from building founda-tions or other vulnerable structures.[15] Where subsurface conditions are suitable there may be infiltration; however, rain gardens can still play a valuable role in locations where impermeable soils occur (like clay for example). Even without the benefit of infiltration, peak flows can be delayed and substantial volumes of water stored in the amended soil and lost through evapotranspiration.

The Idea Spreads

Since 1990s the rain garden concept has spread through the city halls of the United States and has become increasingly well known amongst professionals; however, another fascinating aspect of the growth of interest in rain gardens in the United States has been the way in which citizens have been encouraged to participate in the process. Drainage is something that people tend to leave to the authorities and is only in the forefront of the mind of the average citizen when flooding occurs. In many cities in the United States there are ordinances, which place responsibilities for sidewalk and verge maintenance on the owners of adjacent properties.[16] This situation results in private landowners taking an interest in drainage and having interesting conservations with local authorities about the practicalities of balancing public and private interests. This has led to offers of city authorities offering to take over the maintenance of verges if there is agreement for roadside 'tree lawns' to be converted to rain gardens. Public participation in some American cities in drainage issues now goes beyond this. A good example is in Philadelphia, where the design and installation of a rain garden in Vernon Park, in 2011, brought together an extraordinary partnership of officials, experts, local residents, schools, local clubs, societies and associations, with 100 volunteers involved in the actual installation.[17] This kind of project has taken the creation of rain gardens to a new level, where people work together to improve the management of urban catchments and begin to look for new opportunities to increase the functionality of the open spaces that they see around them.

Some paving in cities is necessary (although there are plenty of places where paving could be removed without inconvenience). Heavily used paths need to be durable. Materials like natural stone and concrete provide the required durability; however, they are impermeable, and rapidly direct polluted water to the drains. This problem can be alleviated through the use of permeable paving. The most commonly encountered permeable paving types are various systems of proprietary interlocking concrete or brick blocks. Water can pass between the blocks, through a sand bed and into a permeable sub-base, where pollutants can be filtered and broken down. Other less common forms of permeable paving include some resin-bound gravels, porous concretes and porous macadam surfaces. Field trials by the United States Environmental Protection Agency in 2000 have shown that permeable paving systems are effective at reducing surface runoff volumes and delaying the peak discharge.[18]

Other Permeable Load-Bearing Surfaces

Where there is a requirement for occasional vehicle parking or access for fire or rescue vehicles, there are a number of techniques, which provide the necessary load-bearing characteristics, whilst also being free draining and capable of supporting vegetation. Best known are the various reinforced grass systems, usually formed from concrete blocks, which include openings or pockets of soil, which can be seeded. Similar systems include those featuring plastic cells or mesh. A drawback with

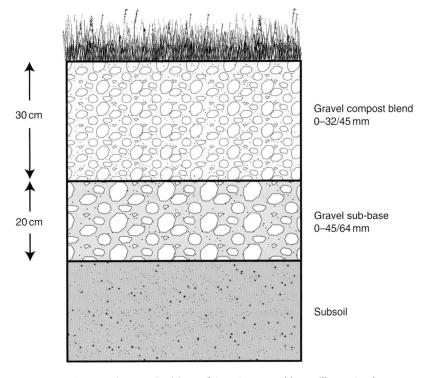

30 cm

20 cm

Gravel compost blend
0–32/45 mm

Gravel sub-base
0–45/64 mm

Subsoil

Figure 9.4 Section showing build up of Austrian gravel lawn. Illustration by Marianna Magklara based on an original by BOKU.

most of these systems is that they are more concrete or plastic than grass, which can become more evident as they wear. Given that the overall approach being described here is to create drainage that mimics nature, any approach that obviates the need for concrete or plastics is to be preferred. In the case of reinforced grass, it is possible to install layers of blended aggregates that can be seeded with suitable grasses or mixtures of grasses and wild flowers. Like any permeable paving, these Austrian gravel lawns (*schotterrasen* in German) need to be installed carefully in order to function correctly; however, specialists at the Institute of Bioengineering and Landscape in Vienna have shown that these buildups can take the weight of heavy goods vehicles whilst supporting a relatively biodiverse acid-grassland sward.[19] If Austrian gravel lawns wear through overuse, they reveal gravel rather than concrete or plastic and can revegetate when rested.

Underground Voids

Once storm water has passed through permeable (or impermeable) paving, it can enter purpose-made voids designed to provide temporary underground storage, for slow release or infiltration, if the geology is suitable. The most common method of creating these underground detention tanks is to use modular polypropylene latticework boxes, which are wrapped in geotextiles to keep out sediments. These modules (usually called cells) can often be stacked to depths of 5 m (16 feet) or more (depending on the particular product) and can take the weight of heavy goods vehicles. Engineers value the predictability of these systems – storage volumes are easy to calculate and performance is relatively easy to model – however, an overreliance on underground storage in drainage design can lead to a neglect of above-ground more natural, multifunctional techniques, especially when there may be competition for space above ground.

Trees and Water

Underground polypropylene boxes are also being increasingly used to provide better underground conditions for trees. The potential for tree pits to receive storm water has been largely overlooked. Arboriculturalists and drainage engineers have had very little to say to each other in the past it seems. There is a common problem for street trees to suffer from

Figure 9.5 Underground space can be created for tree roots and water storage. Illustration by Marianna Magklara.

a lack of water and deliberately directing water toward tree roots can be a useful way of solving this problem and attenuating storm water. Typically street trees are planted into small pits. Often the surface of the pit is sealed or the soil compacted. As young trees grow, roots soon leave the tree pit. Once roots find moisture, they grow rapidly to exploit the source, which could include a break or joint in a drain.[20] Once the roots have made connection with a drain, further growth will be stimulated to allow the tree to take full advantage of the ready supply of water and nutrients. Improving underground conditions for trees can include the provision of purpose-made free draining yet water-absorbent soil, with compaction prevented by the use of polypropylene latticework boxes (soil cells), which can be integrated into the drainage system and can support paving if required.[21]

Stockholm Tree Pits

The doctrine of mimicking nature and minimizing the unnecessary use of manufactured materials like plastics means that we should also consider alternative ways of creating large, free-draining tree pits that can withstand compaction and support pavements. This has been the approach in Stockholm and a few other cities in Scandinavia and Germany. Working for the City of Stockholm, Örjan Stål had observed that trees roots often thrive in rocky ground in contrast to topsoil, which can become compacted and hostile to tree roots. More than 500 struggling trees in the city were revitalized over a 13-year period by replacing compacted soil with rocky substrate. The method involves vacuuming away stale soil and replacing it with consolidated layers of rocks (each rock being between 100 mm and 150 mm across) and to wash a new soil into the voids between the rocks.[22] This buildup provides the necessary structure for heavy-duty paving above or can be left open for planting as required. Although the emphasis has been on improving the health of trees, these tree pits are extremely permeable. Although the rock itself may not be water absorbent, the approach allows for the possibility of creating very large tree pits in streets, with the water-absorbent voids between the rocks able to store significant volumes of water. This means that Stockholm tree pits should also be considered as a potential component of any urban near-natural drainage system.

Conveyance

The various components described so far in this chapter, including green roofs, rain gardens, gravel lawns and carefully constructed tree pits, will absorb most rainfall. It will be infiltrated, stored in the soil or returned to the atmosphere through evapotranspiration, however with very heavy or persistent rain, as source control features become saturated, there will be a requirement to carry water away from a particular place. These so-called conveyance structures should, wherever feasible, be on the surface. Underground pipes or converts should be a last resort. The most common type of conveyance feature is a swale, which often sits on the shoulder of a road or path. In the simplest form these are shallow ditches. Where ground conditions permit, they should allow some infiltration. They can be backfilled with permeable material and may feature a buried underdrain formed by a permeable pipe. Swales tend to be grassy, perhaps to allow easier collection of litter; however there is no reason why they should be not vegetated by other more interesting combinations of perennial vegetation. It is commonplace for swales to include a series of check dams, which can increase infiltration and have the effect of creating small ponds or wetlands. This means that swales have the potential to provide a wide range of valuable habitats from bare ground and dry grasslands through to marshes and ponds.

Rills

Where space is limited in the inner city it may still be possible to provide conveyance in an open channel or rill. Although channels lined with waterproof material (typically stone but brick, concrete and even iron and steel might be used) do not increase infiltration, they present an opportunity for the urban designer to adorn and animate the street. Where channels must be protected by grilles, for practical reasons (to prevent trips and slips or to collect trash for example), they can also include unique patterns, that advertise and celebrate the water cycle.

Ponds

Swales and rills might discharge into conventional drains or watercourses, however, where space permits or regulations require, before this they may discharge into small-scale local basins or ponds or, in some cases, much larger basins or ponds, which constitute so-called regional control features. Basins and ponds provide additional attenuation, storage and cleansing of surface waters.

Detention Ponds

A detention pond is a basin that would normally be dry but which can be filled with water during floods. The landform would normally be some kind of basin with a throttled outlet to slow outflows. Inlets may be set at a particular level, which means that they function only when the drainage system reaches a particular flow threshold. Detention basins are often grassy open spaces, which are used for informal recreation in everyday dry weather.

Attenuation Ponds

An attenuation pond is a water body designed to temporarily fill with water during wet weather. They are usually wet but have extra depth to allow storage of large volumes of water, which are relatively easy to calculate and model. The rapid increase in water depth, which can occur following heavy rain (especially if upstream attenuation is lacking), means that there is often a perception of danger with attenuation ponds. It is commonplace to find attenuation ponds that are fenced and this is often done in a way that makes these ponds unattractive or even foreboding in appearance. Attenuation ponds also tend to be depositories of sediment. Again if upstream collection of silt is inadequate, ponds can quickly fill with silt and this can lead to higher maintenance costs. On the positive side, attenuation ponds protect watercourses from the ingress of smothering sediments and damaging pollutants. Ponds can include marginal aquatic vegetation – a valuable habitat itself – designed to maximize the capability of ponds to filter and clean water.

Floating Wetlands

One of the difficulties with the intermittent nature of flood and the fluctuating levels of water in attenuation ponds is that with many marginal areas, wetland and aquatic vegetation may not persist and the important

Figure 9.6 Wetland raft. Illustration by Marianna Magklara.

habitat value and cleaning function associated with that element is lost. Although rarely used, one solution to this problem is floating wetlands. Rafts can be created and vegetated with wetland vegetation.[23] These rafts can be tethered and will rise and fall as water levels in a pond fluctuate, continuing to support biofilms of algae and microbes on the roots that sequester dissolved pollutants, including nitrates and phosphates. Since 2009, the City of Billings, Montana, for example, has successfully used floating wetlands to improve the quality of water in a pond receiving runoff from vehicle parking areas in an industrial estate.[24] The undersides of rafts can provide useful refuges for fish and aquatic invertebrates and the vegetation above, feeding and nesting places for waterfowl.

Larger Water Bodies

If the sustainable drainage management train philosophy is conscientiously applied, so that source control methods are maximized, the need for large water bodies to attenuate water can normally be avoided; however, if there are large areas of hardstanding that cannot be broken up or large areas of roofs that cannot be greened, if the underlying rock is impermeable or the topography within the catchment is liable to funnel sudden pulses of water through urban areas, there may the need for larger water bodies – so called regional control features. An example of large water bodies catching floodwaters in urban areas is in Ljubljana, Slovenia. This urban region is vulnerable to destructive floods caused by a number of torrents, which are fed by intensive rainstorms or snowmelt in the upper catchment. The region also suffers from occasions when groundwater rises through the karst. By 1851, the urban rivers had been modified and people living in the Ljubljana Marshes to the southwest of the city, were being advised by the authorities to raise the ground levels of their buildings. The city-region suffered six severe floods during the twentieth century, and particularly severe consequences in 2010, when heavy rainfall caused several fatalities and millions of euros of damage. The Podutik reservoir was created in 1986 on the Glinščica River to protect nearby settlements by storing floodwaters. The reservoir is performing adequately in terms of its ability to store floodwaters; however, inflows from polluted tributaries, urban runoff and leaking septic tanks have caused water quality problems. Since 2006, the authorities have addressed this problem by using green infrastructure in the form of vegetated ditches to improve water quality.[25] This has improved the nature conservation value of the reservoir and also encouraged recreational and educational use, so that the project has become an example of multifunctional, near-natural drainage.

This review shows that near-natural urban drainage can involve a whole host of interventions, along the whole journey taken by water, from the rooftops to the river. There should not be an overreliance on large water bodies, with water being captured, evaporated or infiltrated across the whole catchment, on green roofs, rain gardens, tree trenches and swales. Given the extreme weather that can occur and which is predicted to become more frequent, cities may have to also look at ways of leaving open space where large volumes of water may be temporarily stored. When planning new settlements, this will mean leaving more space for green infrastructure, particularly in the most flood-prone areas; however, with existing settlements, especially those that have been hurriedly and thoughtlessly constructed in floodplains and other vulnerable areas, homes and places of work may need to be moved to safer areas – sooner with good planning and support or later after catastrophic flooding is repeated – before communities have had time to recover. The near-natural approach means leaving more space for water and nature in urban areas.

10. Reduce

A Worthwhile Effort

Water is too important and too valuable a resource to use extravagantly. Reductions in consumption are necessary and possible. Some of the initiatives and suggestions described in this chapter may seem small scale and the volumes cited may be small in terms of the overall volumes of water consumed by civilization in its entirety; however, many modest savings can combine to make substantial overall savings, which will make a real difference. Water conservation should become an issue that everyone is aware of and takes responsibility for, with constant attention being paid to the issue. The prevailing attitude is to disregard water conservation until a crisis occurs. A cultural change is required with people constantly seeking new ways of conserving water. So what can be done to reduce consumption?

Reduce Leaks

Many cities still need to do more to reduce leaks in the water distribution network. Tokyo has much to teach others. That city of 12 million people reduced losses from 150 million cubic metres per year in 2000 to 68 million cubic metres a decade later. The amount saved is equivalent to about 14 days' consumption.[1] Changes that have enabled Tokyo's Bureau of Waterworks to make these savings include continuous computer controlled monitoring and flows (and therefore leaks) and a new culture where pipes are replaced as soon as leaks are detected.

The Water Sensitive City, First Edition. Gary Grant.
© 2016 John Wiley & Sons, Ltd. Published 2016 by John Wiley & Sons, Ltd.

Monitor

The first task for the householder or business owner is to measure consumption. Nowadays in most domestic and commercial premises, water is metered. Usually total water consumption in cubic metres (or litres) is measured and, occasionally, rates of flow are also recorded. The most commonly encountered meters incorporate oscillating pistons or rotating discs. Less common are meters that send jets of water against impellers. Where large volumes of water are being monitored for water distribution or commercial use, turbine meters are used. Mechanical meters are usually protected from small stones and other debris by strainers. Other less common categories of meters are the magnetic flow meters that use electromagnets to measure water flow velocities and the ultrasonic meters, which can measure flow rates by measuring changes in the velocity of ultrasonic sound waves as they pass through the water. Water meters are usually installed for billing purposes; however, if you live in a place where water is not metered you may have the option of requesting a meter. In the United Kingdom, for example (where about 40% of properties have meters), OFWAT, the regulator of the water services industry, provides advice on the procedure for requesting water metering.[2] There are also inexpensive meters available which will enable you to measure water consumption of particular installations (like washing machines) or operations (like garden irrigation for example).

Figure 10.1 Domestic water meter. Photograph by Jack Zhou.

Check for Leaks

Once meters are installed these can be used to test for leaks. Taps and appliances are then switched off (with the exception of the main shut-off valve or stopcock). If the meter continues to record flows, after say, 30 minutes, that is an indication of a leak. In order to narrow down the location of a leak it may be possible to turn off isolation valves until the flow through the meter ceases. Then it may be necessary to make visual inspections, beginning with the meter itself and moving on from the main shut-off valve (stopcock) and then working through the system. The most common place for water to be leaking in a house is the toilet, where water can leak from the cistern straight into the toilet bowl or through an overflow pipe. Leaks that are not obvious (like an overflow into a toilet bowl) can be traced using food colourants. It is also important to check taps and showerheads for drips, which can often be easily fixed by replacing washers or other inexpensive components.

Less Flush

Once leaks are fixed, the water saving effort can move on to the modification or replacement of equipment. About a third of all potable water used in the household is used for toilet flushing. With older toilets installed before 2001, it is possible to reduce the amount of water used in each flush by between 1 and 3 litres by using a cistern displacement device (CDD).[3] These are usually water-filled bags, which are sometimes provided free of charge by water companies. People have also occasionally used bricks or pebbles as CDDs, although care must be taken to avoid interfering with the operation of the moving parts in the cistern. In a typical house where the toilet is flushed 5000 times a year, a simple intervention like this would save between 5000 and 15,000 litres of water during that period. When the time comes to replace the toilet, it should be replaced with a more efficient model. New dual flush toilets, which have a split button to provide a choice of flush volumes, may use a little as 4 litres per flush – less than a third of the 13 litres used by some of the old-fashioned toilet cisterns. Manufacturers claim that a family of four can save 17,000 litres a year by using the economy button on a dual-flush, water-efficient toilet.[4]

Toilets are Not for Trash

Much water is wasted by people flushing away trash, including cotton buds, cotton wool and cleaning wipes (also known as wet wipes or towelettes). Items like this should be disposed of very carefully.

Not only is water wasted when these materials are flushed, but sewers and pumps in wastewater treatment works are being blocked. In 2011, Thames Water, which is responsible for sewage treatment in the London area, was getting 2000 calls a month to remove cleaning wipes from drains. This operation costs £12 million a year.[5] Equipment is required to remove cleaning wipes from sewage as it enters sewage treatment works and energy, time and money is expended by the sewage treatment companies sending these materials to landfill.

Composting Toilets

It is possible to have composting toilets that work entirely without water.[6] These toilets are usually installed in locations where a water supply and sewers are not available. Although it would make sense in terms of water conservation to make more use of this technology, in practice sewage systems all over the world are planned and designed to receive waste from water closets and this is set to continue. It is unlikely that composting toilets will become commonplace in cities in the foreseeable future.

Showers

Showering may use up to a fifth of the water consumed in a typical household. More people than ever use a shower and it has been suggested that people are tending to spend longer in the shower. In 1999, the average American shower used 65 litres and lasted for just over 8 minutes.[7] There is also much effort given to saving water used in showers because of the energy used in heating water. It is possible to select shower heads that simply reduce flows or mix the water with air to reduce flows, typically by 50%.[8] WaterSense is a water conservation partnership launched by the United States Environmental Protection Agency in 2006. In 2009, WaterSense released specifications for shower heads, with a rigorous set of performance requirements for force, cover and flow. Flows for shower heads, according to this specification, should not exceed 2 gallons per minute (7.6 litres per minute).[9] Another water-saving strategy is to spend less time in the shower. Purpose-made timers are available to help people keep track of the time they spend in the shower, with 4 minutes thought to be long enough.[10]

Washing Machines

In the United Kingdom, washing machines use about 15% of the water used in homes. There is some variation in the amount of water used by machines, with the 8 kg capacity machines found in the typical home using as much as 14.1 litres/kg and as little as 5.5 litres/kg. There are reports of the most water-efficient machines being unsatisfactory in terms of cleaning, leaving visible traces of detergent; however, there are machines that provide a good balance between cleaning capability and water conservation, using fewer than 7.5 litres/kg.[11] Machines should be fully loaded to ensure energy and water efficiency and the selected programme should be matched to the fabric type – water consumption can vary considerably between the various programmes provided by the typical washing machine.

Dishwashers

Like washing machines, the water consumption of dishwashers can also vary considerably. It is recommended that dishwashers are fully loaded and plates are not prerinsed since this is unnecessary and therefore wastes water. Many modern dishwashing machines have 'eco' or 'economy' settings designed to save energy and water. The flow volumes of kitchen taps also vary, from 2 litres, up to 25 litres per minute. Using a kitchen bowl or inserting the plug in the sink can reduce the amount of water used to wash vegetables by 50%. In the United Kingdom, the amount of water used for kitchen taps and dishwashers represents between 8% and 14% of the water consumed in the home.[12] Relatively little water is used for drinking; however, in order to save running a tap until the water is cold and wasting 10 litres, it is possible to fill large bottles and leave them in the fridge.

Garden Irrigation

Households also use water in the garden. Potable water should never be used to irrigate gardens. There is no necessity to irrigate domestic lawns. Most activities can continue on a brown lawn and they quickly recover their green colour after a dry spell. Many lawns are unused and can easily be converted into xerophytic plots with drought-tolerant vegetation or rain gardens that are fed by surface water runoff (see Chapter 9). A typical lawn sprinkler uses 10,000 litres per hour, which is more than an entire household might use in a day. If irrigation is absolutely necessary to keep particular plantings alive, then irrigation water should be provided by collecting rainwater (see Chapter 11) or recycling greywater (see Chapter 12). For many people, pressure washing of vehicles and paving is more of a pastime than a necessity. Pressure washing should be avoided wherever possible, but where its use is considered to be essential, a water and energy-efficient model should be used.

The Workplace

Cities are more than domestic dwellings and plenty of water is consumed in the workplace and in commercial premises. Every workplace should promote water conservation through staff meetings and the circulation of information and advice. Volunteers should be recruited to become water conservation champions, who can monitor water use and suggest targets for saving water. Reference to industry benchmarks and initiatives in similar situations can help to ensure that these targets are realistic and achievable. Most of the advice relating to households also applies to the workplace; however, there may be opportunities to save relatively large quantities of water with the larger scale of operations in the commercial sector. Even simple interventions like changing taps (for example infrared-controlled taps with timers) and shower heads can lead to significant savings. For example the Holiday Inn in Flinders, Australia, was able to cut its water use by 50% and to recoup its AUD$ 22,000 investment in new fittings in 18 months.[13] According to England's Environment Agency, urinal flushing can represent most of the water used in many public and commercial buildings. Chesswood Middle School reduced its annual water consumption by 68% (a saving of 900 m^3) by installing controls. An investment of £960 saved £1414 in the first year (a payback period of 8 months). Urinal controls can include timers, restricting flushing to certain periods or when people are using a building, or can involve the use of infrared movement sensors or switches activated by the opening or closing of doors. There are also waterless urinals, which as well as saving water, do not suffer from the problems of blockage and flooding which often occur with conventional urinals.[14]

Behaviour Change

Another way of saving water is to modify behaviour and patterns of consumption. For example, hotels will often suggest that towels and linen are not changed and washed every day. Some guests object to this kind of initiative, believing it to be motivated by a desire to reduce costs rather than a genuine effort to save water. Hotels have overcome this difficulty by allowing guests to opt out if they wish, or in some cases have engaged guests by offering incentives. The Starwood Hotel group, for example, offers guests gift shop vouchers worth $5 or loyalty points if they don't have their room (and therefore linen and towels) cleaned every day.[15]

Heating, Ventilation and Air Conditioning

Many retailers, warehouses, schools, museums, offices and factories use heating, ventilation and air conditioning systems (HVAC) to provide comfortable conditions for staff and visitors or, in some cases, special conditions required for manufacturing processes or safe storage. Figures vary considerably according to building types and sizes and the severity of the exterior climate but air conditioning and heating can account for between 10% and 25% of a building's water consumption.[16] Most HVAC systems are designed as closed loops where water is recirculated and relatively little is consumed; however, cooling towers are the places where water is lost from these systems. Most losses are through evaporation and some through drift (a process where mist is caught by the wind). Water is also lost from HVAC systems when water is drained off as part of maintenance or cleaning operations. It is possible to use existing building system management software to control cooling tower operations. Water saving measures might include the automatic shutting down of cooling towers whenever it is possible to do so without reducing the comfort of a building's occupants. There are several other ways in which HVAC can be improved and operated in order to save water. Water levels in cooling towers can be more carefully monitored to prevent overflow. Alarms and more efficient valves can be installed. Elements can be added to redirect airflow and reduce water loss caused by drift. Evaporation rates increase with cooling loads, so any interventions that reduce the need for cooling (like white roofs or green roofs for example) will save energy as well as water. Water quality within systems can be more carefully monitored, reducing the need for water to be bled from systems. Initiatives like these can also save money. A life-cycle cost study commissioned by the California Statewide Utility Codes and Standards Program in 2011 showed that it would cost $3500 to install a particular package of cooling tower improvements at a typical office building and operate them for 15 years. The measures described would provide more than $11,000 in water and water treatment savings during that same time period. The same study advises that even greater returns on investment are achievable in buildings that use more water, including schools, hospitals, or factories.[17]

Vehicle Washing

There are concerns over vehicle washing, both in terms of the pollution of ground water but also the quantities of potable water that can be used. The effect of vehicle washing on water quality is discussed in Chapter 13. People wash their own cars in the street or driveway but cities also have fleets of buses, heavy goods vehicles, industrial vehicles and trains, which are all washed, sometimes in yards with simple drains or in more sophisticated purpose-built facilities. The amount of water used to wash a single car can vary between 30 and 100 US gallons (113 and 378 litres).[18] The Alliance for Water Efficiency has set a target of using 40 US gallons (150 litres) of water or less to wash a small vehicle and detailed advice is available on how to do this.[19] For example it is possible to use a fixed volume of water in a bucket rather than a continuously running hose. Purpose-built facilities are also available that clean and recycle water (see Chapter 12). The total quantity of water (and energy) used in vehicle washing can also be significantly reduced by reducing the total number of times a vehicle is washed. Instead of washing vehicles according to a fixed routine, irrespective of how dirty or clean they are, they can instead, be washed as the need arises.

Urban Farming and Recycled Water

As discussed in Chapter 3 (Demand), there are many products and services consumed in towns and cities, which involve the use of large volumes of water in their production or delivery. Given that agriculture consumes around 70% of total global water withdrawals (a figure that rises to more than 80% in arid regions) it would be worth considering how those huge volumes can reduced. For the most part, urban dwellers are not farmers; however, cities are a plentiful source of wastewater (in particular grey water), which could easily be recycled to grow some food and therefore reduce the amount of water abstracted from the wider environment. Although some progress will be made in increasing the amount of food grown locally within cities, cities will continue to rely on food produced by farmers elsewhere. Urban dwellers can really influence rates of water abstraction through their choice of diet, for some foods require much more water in their production than others.

Diet and Water

The trend for societies to increase their consumption of meat as incomes rise has a profound effect on water withdrawals because, in general, the production of most species of meat requires more water than the production of vegetable foodstuffs. According to the Water Footprint Network, the production of 1 kg of beef requires 15,415 litres of water. Most of this water is required for the irrigation of the grain and grass required for cattle feed. For each kilogram of beef produced an animal must consume more than 40 times that weight of vegetation. By contrast the production of 1 kg of vegetables requires 322 litres of water.[20]

Soft Drinks

The production of soft drinks is particularly wasteful of water. The production of a litre of a sugary carbonated soft drink varies considerably from region to region, but can use more than 600 litres of water.

The production of soft drinks requires several steps, including the production of bottles or cans, cleaning, filling and packing; however, most water consumption is associated with the production of the ingredients, including high-fructose maize syrup (or sugar cane or sugar beet).[21] Products like sugary carbonated soft drinks are probably best avoided, given their enormous water footprints and the increased risk of tooth decay and contracting type-2 diabetes and heart disease.[22]

Clothing

Another category of water guzzling consumables is clothing. Cotton is the most common natural textile used in the manufacture of clothing. Cotton is often grown in arid areas where irrigation water is abstracted from rivers, which become ecologically degraded. A cotton t-shirt has a water footprint of 2,495 litres and a pair of jeans a water footprint of 9,982 litres.[23] Buying fewer items of clothing and wearing out old clothes before replacing them will save water. There are several initiatives to manufacture cloth from recycled waste materials including food and plastics, initiatives that will reduce waste and save energy and water.[24] There are also some clothing manufacturers who are working with farmers to reduce the water footprint of products. Therefore changes in consumer behaviour can have an impact in reducing water consumption in places far from the point of purchase.

Reduction Targets

Reduction in the consumption of water by itself will not create the water-sensitive city and will not solve all of the ecological problems experienced in the watersheds that supply the foods and products consumed in cities; however, reductions will make a difference as part of an integrated approach, which includes collection (see Chapter 12) and recycling (see Chapter 13). In many cities, overall savings in consumption of 50% are achievable through the reduction of leaks in the distribution system combined with other measures. A reduction of 50% of the water consumed within homes and business is also possible through the installation of water-efficient fittings. In some cities there is profligate use of water for the irrigation of ornamental gardens, which is, arguably, completely unnecessary. Therefore there is further scope for savings – it should be possible for per capita daily domestic consumption to fall from the more than 500 litres per day recorded in many American cities to the 80 litres per capita per day used in Chinese cities.[25]

11. Collect

Reduce Reliance on Abstraction

The collection and use of rainwater reduces our reliance on water from watercourses or underground aquifers. In the State of Tamil Nadu in India, it was a concern over the depletion of groundwater caused by overabstraction that motivated the legislature to make rainwater harvesting compulsory in 2001.[1] There are already welcome signs of progress, with reports that groundwater quality has improved and quantities of groundwater are increasing, even though compliance with the new legislation is patchy. This initiative in Tamil Nadu has also become a model for other states in India and elsewhere in developing countries.[2] Nearby Sri Lanka, for example, passed legislation in 2007 to promote rainwater harvesting following the establishment of a forum a decade earlier, in 1997. The forum continues to promote and provide advice to support rainwater harvesting.[3]

When Sealed Surfaces are Useful

The preponderance of sealed surfaces in towns and cities may be a problem in the sense that it causes rapid runoff, which tends to exacerbate flooding, but it is also an opportunity for those wanting to collect and store water. Conventional roofs are especially convenient collectors, with waterproofing that does not usually taint water, and downpipes which can easily be used to redirect water to storage tanks rather than drains.

The Water Sensitive City, First Edition. Gary Grant.
© 2016 John Wiley & Sons, Ltd. Published 2016 by John Wiley & Sons, Ltd.

Rainwater Harvesting

Rainwater harvesting, as it is called, usually involves the collection of water from roofs and storage in tanks for subsequent use. For useful quantities of water to be collected, large roofs are preferred and the capacity of storage tanks needs to be sufficient to store enough water to provide adequate supplies during dry spells. Where there is a reliance on harvested rainwater, very large tanks may be required. In most situations harvested rainfall supplements other supplies, perhaps providing a backup source during times when piped supplies are suspended or when there is a drought. If annual rainfall is less than 400 mm, rainfall harvesting may not be feasible.[4] Even when piped supplies are reliable, use of rainwater can help to reduce consumption and therefore extraction, from rivers and aquifers and in most jurisdictions, save money. Rainwater is relatively clean and can be readily treated to make it potable and large-scale rainwater harvesting could have a role to play in reducing floods.

How Rainwater is Tainted

Although rainwater is relatively clean and free of salts, it can be acidic as it dissolves atmospheric carbon dioxide or pollutants like sulphur dioxide. However, once it lands on a roof or other sealed surface it can pick up pollutants, including faecal matter from animals and birds, plant

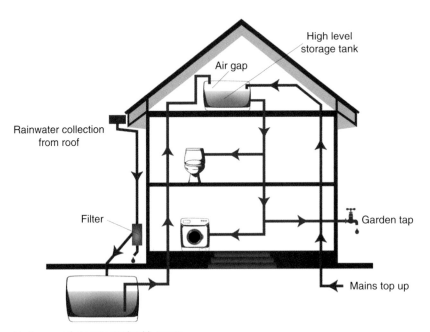

Underground storage tank with pump

Figure 11.1 Typical domestic rainwater harvesting scheme. Illustration by Marianna Magklara.

debris and dust. Also some roofing materials may decompose or corrode, causing particles to enter the water, including metal oxides. Roofs made from smooth materials like metals result in the smallest losses. There are metal roofing products, including steel sheets, which have been coated with a mixture of aluminium and zinc, which are considered ideal for rainwater harvesting.[5] Slate is another smooth material, traditionally used for roof tiles, which is ideal for collecting rainwater. Concrete and clay tiles are inert, but slightly porous and can result in evaporative losses. Some waterproofing materials, like asphalt, are considered unsuitable for collecting rainwater that is to be subsequently used for drinking because of the leaching of toxins; however, water collected from such roofs can be used for irrigation without problems.[6] It is occasionally argued that rainwater should not be harvested from green roofs. Although about 50% of the rain that falls on an extensive green roof is lost through evaporation over the course of a year, water does run off green roofs following persistent rain. The water that comes from a green roof is usually yellow in colour (the tint is caused by coloured dissolved organic matter, sometime shortened to CDOM) and not suitable for drinking water supply, but it can be used for irrigation or toilet flushing. Where it is known that a green roof is to be used as a rainwater-harvesting surface, substrates and filters can be specified that reduce the amount of CDOM entering downpipes. An example of a commercial building with a green roof where rainwater is harvested is the Adnams distribution centre in Suffolk, England, where, in a region that has 550 mm of rain per year, 650,000 litres of water is collected from a 4000 m² green roof, suggesting a runoff coefficient of 30%.[7]

First Flush

The first flush of runoff following rain is the most polluted and therefore many rainwater harvesting systems divert the first few dozen litres to the drain (or better still a rain garden) before channelling water to storage tanks. First-flush diverters are usually formed from blanked-off vertical pipes that fill up and overflow to an outlet, leaving sediments and pollutants in the bottom of the pipe. The amount of first flush that should be diverted varies considerably depending on how much dust has accumulated on a roof. The Texas Manual on Rainwater Harvesting recommends that the first 50 litres of flow from every 100 m² of roof should be diverted away from the cistern.

Novel Methods

A concern over the quality of runoff from roofs and losses of first-flush flows diverted to avoid contamination has led people to propose alternative ways of collecting rainwater. An example of this is the RainSaucer, which looks like an upside-down umbrella. It consists of a food-grade polythene dish connected to a filter and sealed container. This product is being promoted as a way in which people can reduce their reliance on expensive bottled water, especially in some developing countries where there are suspicions that municipal water supplies may be unhealthy. These gadgets are certainly cost-effective in countries like Guatemala, where many people spend $300 on bottled drinking water each year – the equivalent of one month's salary.[8]

Filters and Tanks

Most harvested rainwater, however, will continue to be collected from roofs. In order to keep that water clean, as it leaves the roof of a typical rainwater harvesting system to enter a cistern it is usually fed through a filter tank (sometimes known as a roof washer), usually a modestly sized tank (with a capacity of around 200 litres), which has a coarse strainer to remove larger items like leaves but also a fine mesh filter (typically 30 microns). Filters tanks need to be regularly cleaned in order to prevent the buildup of sludge. Once filtered, water can enter the main storage tank. The size of a tank is determined by the demand, the length of dry spells between rain and how much rain falls. The storage tank is usually the most expensive part of the system and so affordability is an important consideration. In some very urban situations where space is limited, finding somewhere to locate a water-storage tank can be difficult, but once the value and importance of rainwater harvesting is understood, space can usually be found. Storage tanks or cisterns have been made from many different materials over the centuries; however, there are a few basic requirements. Storage tanks must be opaque. If light enters the tank it will encourage the growth of algae. For tanks that will be used to store potable water, old tanks that may have been previously used to store materials that contain or could leach toxins must be avoided. Tanks must be covered to prevent creatures entering, particularly insects like mosquitoes and other flies that like to lay their eggs in water. Finally, water-storage tanks should be accessible for cleaning and maintenance.

Siting a Tank

In order to be gravity fed, storage tanks are usually located at a lower level than the surface from which water is collected. They should be close to the point of collection and also point of use if possible, to minimize the distance that water is conveyed or pumped. Ideally a tank will be higher than the point of consumption to avoid or minimize the need for pumping. In most situations, however, especially in level terrain, pumps are required to move water to header tanks or pressurize water supplies. Sometimes siting a tank near the main point of supply means that the amount of new pipework required can be minimized. Overflows of tanks should not be routed in a way that could undermine foundations or flood nearby septic tanks or sewers. Most authorities recommend that overflow is directed into a drain or at least 3 m (10 feet) away from foundations if entering the soil directly. It is also important that tanks are protected from runoff from stables or wastewater treatment facilities. Water is heavy (it weighs 1 kg per litre) and therefore water tanks require adequate, level and secure footings.

Materials

Water-storage tanks have been made from a wide variety of materials. Fibreglass is durable but relatively expensive and tends to be more widely used for large-capacity tanks. Polypropylene and polyethylene tanks are relatively cheap and therefore are the most popular materials for small storage tanks; however, they must be protected from ultraviolet light, which degrades the plastic. It may also be necessary to reinforce polypropylene and polyethylene tanks with metal cages. Metal tanks were popular before the plastic age; however, they are relatively expensive and prone to corrosion and are therefore used less frequently nowadays. In many cases metal tanks are lined with plastic. Concrete and masonry may be suitable for larger projects, however these materials are vulnerable to cracking and can be expensive to maintain. Certain species of wood are durable when wet and suitable for use in water-storage tanks. Wooden tanks are easy to maintain and there may be aesthetic benefits (especially when tanks are on view) however wooden tanks are expensive to build.

Treating Rainwater

For rainwater that is to be used for irrigation or toilet flushing, little treatment is required. Coarse filters can remove large items and smaller filters sediments that could cause clogging of small pipes and emitters. For potable water, however, it is important that pathogens are removed. In most urban situations where a reliable potable water supply is available it would not make economic sense to treat water to such high standards; however, in rural locations where rainwater is treated to drinking-water standard this usually involves fine fibre and activated charcoal filtration followed by irradiation with ultraviolet light. Other less common methods of treatment include sand filters, ozone, reverse osmosis and chlorination. Pathogens, which could occur in contaminated rainwater, include coliform bacteria (particularly faecal coliforms and *Escherichia coli*),[9] *Cryptosporidium* and *Giardia lamblia* (protozoans that cause gastrointestinal illness),[10] *Legionella pneumophila* (a bacterium which thrives in warm water)[11] and viruses. Water that is intended for drinking should be tested regularly to ensure that it is not contaminated.

Sizing Tanks

Designers of rainwater harvesting systems try to match supply (the amount of rainwater that can be captured and stored) to demand (the volume of water required). Usually the process begins with an evaluation of intended use and predictions of the volume of water that will be required. Designers should ask: 'What are the daily, monthly and seasonal demands?' 'How much water will need to be stored and how much can be captured and when?' Potentially a litre of water can be collected for every millimetre of rain that falls on every square metre of sealed surface; however, there are losses from splashes and evaporation as well as further losses caused by the diversion of the first flush and filtration. The 'collection efficiency' or 'runoff coefficient' of the best metal roofs can be as high as 95% and a typical figure used to estimate collection efficiency in United States examples is 90%. The Environment Agency in England suggests using a figure of 80% for the runoff coefficient of farm building roofs in the absence of more specific data.[12] Most rainwater harvesters assume losses of between 10% and 25% of rainfall falling on collection surfaces depending on the materials used, temperature and humidity (which affects evaporation rates), first flush diversion and filtration. Very intense rainstorms may increase losses further, overwhelming gutters and filter tanks. Of course, with prolonged rain, most storage tanks will eventually fill and overflow. To ensure long-term supply the catchment area must be sufficiently large and the storage capacity must be sufficiently voluminous to provide water through the longest dry periods, which could be several months in some climates. For example, annual rainfall in Texas varies between 200 mm in the west of the state to 1400 mm in the east. In the dry west, a typical homestead would not be able to collect enough rainfall to water a whole family; however, further east this is possible, providing enough water can be stored during dry spells, which could be for as long as three months.[13] For a family of four using a modest (by American standards) 150 litres of water per day per person, a tank or tanks with a total storage capacity of at least

54,000 litres (54 m³) would be required. A two-storey (say 5 m high) cylindrical tank would need to be 4 m wide to store that volume of water – similar in scale to a sizeable building in a suburban or rural context.

City Centre Rainwater Harvesting

Rainwater harvesting in city centres is also becoming increasingly common, particularly with larger buildings like offices, hotels and schools. Sustainable building assessment and certification systems including BREEAM[14] in the United Kingdom, LEED[15] in the United States and the international scheme for hotels Green Key[16] are encouraging rainwater harvesting schemes in new buildings. An example is Horizon House, in Bristol, which is the headquarters building of England's Environment Agency.[17] Water collected from a 2000 m² roof is stored in a basement tank, which can store 30,000 litres. This water is disinfected using ultraviolet light and then used to flush toilets. Rainwater harvesting, combined with water-saving devices, has helped staff to reduce their consumption of potable water significantly since they moved into the new building in 2010.[18]

Potsdamer Platz

Large urban renewal projects represent an opportunity to capture large volumes of water and to integrate rainwater harvesting fully with landscape design. This means that landscape features can be included that store water for reuse. It also means that it is possible to create precincts that are zero discharge – that is, places where no water leaves the site, except perhaps during the most extreme storms. An example of this is the Potsdamer Platz in Berlin. Once an important junction and plaza, the area was badly damaged during World War II and then divided by the Berlin Wall. The fall of the Berlin Wall in 1989 provided an opportunity to redevelop the area. The new complex of shops, offices and theatres was completed in 1998. The landscape was masterminded by the German firm of landscape architects, Atelier Dreiseitl, which has developed a reputation for a sensitive consideration of water and ecology in its projects.[19] About 525 mm of rain falls on the site each year. Water from both vegetated and unvegetated roofs is channelled to lower roof gardens, with excess from those areas entering a lake above an underground car park. The lake includes plenty of marginal aquatic vegetation, which provides habitat but also helps to maintain good water quality. Water from the lake can be used to irrigate the gardens within the project during dry weather. Rainwater is also stored in five underground cisterns and water from these is channelled in the open air through rills and across the piazza in a series of attractive features, which provide people with opportunities to see and enjoy the water.[20]

District Collection

As well as integrated rainwater management in large developments, it is also possible to collect rainwater from urban districts by channelling small streams into reservoirs. Since 2006, in Royal Park, Melbourne, Australia, stormwater from adjacent urban areas has been diverted into a new storage basin, which can hold 12 megalitres (12 million litres). A little later, this storage basin was supplemented by an underground tank with a capacity of 5 million litres. Only low flows from a minor watercourse are redirected into the storage basin and sediment is trapped in a wetland feature before water overflows into the storage area, which is, in effect, a park lake. Overflows from the basin enter the nearby creek much cleaner than the urban runoff that enters the park's new wetland. Water from the basin and tank is treated with ultraviolet light and used to irrigate the adjacent golf course, sports fields and street trees.[21]

Singapore Wants Every Drop

Singapore is a densely populated city-state without natural aquifers or lakes, with a government that wants to reduce its reliance on water imported from the neighbouring country, Malaysia. The second of two agreements with Malaysia to import water will expire in 2061. This has now become a target date for the city to work towards in an effort to implement a comprehensive and holistic resources management scheme. Singapore's national water agency recognizes that it has to maximize opportunities to harvest rainwater as part of this holistic approach and it does collect urban runoff for water supply on a scale that does not occur anywhere else. Two-thirds of the whole land surface of Singapore acts as a water catchment, channelling rainwater through a network of drains into 17 reservoirs. Already, all of Singapore's estuaries have been dammed to create freshwater storage reservoirs and there are plans to increase the rainwater catchment areas to 90% of Singapore by 2060. This will mean including many of the minor watercourses close to the shoreline, where there are issues with salinity; however, Singapore is confident that it will have the monitoring and technology to deal with this, as well as the other water quality issues associated with capturing and treating urban runoff.[22]

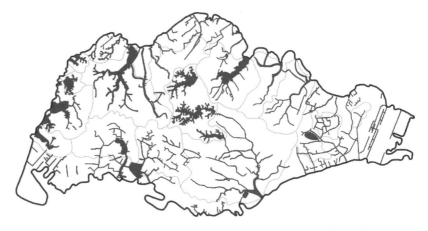

Figure 11.2 Singapore's catchments and reservoirs. Illustration by Marianna Magklara.

A recently created low-lying reservoir is the Marina Barrage, which collects runoff from a 10,000 ha urban area where more than a million people live, creating a challenge for the management of the quality of urban runoff. A 350 m wide barrage separates Marina Bay from the sea, and the new reservoir receives surface water runoff from a number of watercourses including the Geylang River, Kallang River, Pelton Canal, Rochor Canal, Singapore River, Stamford Canal and Sungei Whampoa. Construction of the barrage was completed in 2008 and within 2 years the impoundment had been successfully converted from an estuary to a freshwater reservoir by allowing water to flush through the barrage at low tide. It has been estimated that this reservoir will supply 10% of Singapore's freshwater needs.[23] In order for Singapore to maintain good water quality in its urban reservoirs and to minimize the cost of treatment, various interventions have been necessary, including large scale rain gardens designed to remove nitrogen, phosphorus and suspended solids. An example of a rain garden designed to improve water quality is that on the Balam Estate, a 240 m² feature designed to treat a catchment of 6000 m² using a 300 mm deep bed of rocks and wood chips.[24] Other initiatives in this urban catchment include green roofs, permeable paving, restored wetlands in the watercourses, constructed wetlands and floating wetland planters.

Legal Problems

In parts of the western United States, water rights assigned more than a century ago, designed to prevent ranchers and other landowners from depriving others downstream, mean that it is illegal to harvest rainwater. When car dealer Mark Miller of Salt Lake City, Utah decided to collect rainwater from the roof of his new building and use that water in his new superefficient and intelligent car wash, he was surprised to be informed by state officials that he was breaking the law.[25] Although it would make sense to collect rainwater in parched Salt Lake City, the water rights in that town belong exclusively to various government bodies. Salt Lake City officials did eventually work out a way of permitting Miller to continue to collect rainwater; however, the case has highlighted the anachronistic regulations, which also apply in other states in the West. In 2009, legislators in Colorado passed laws to allow homeowners to collect rainwater and by 2014 the practice was being encouraged in places like Douglas County to the south of Denver.[26]

Dew

In places where there is sufficient moisture in the air, it is possible to collect water by promoting condensation. Dew forms naturally on surfaces when the surface temperatures of solid material fall, usually in the morning. The temperature at which water vapour forms droplets on a solid surface is called the dew point, when the air is saturated with water at any given pressure.[27] The dew point is affected by humidity as well as temperature, so that in places where rainfall is low but humidity is high and surface temperatures fall sufficiently, the collection of dew can be worthwhile.

Lanzarote

On Lanzarote in the Canary Islands, 100 km from the coast of Morocco, where rainfall averages just 140 mm a year, it was noticed, after a series of volcanic eruptions in the 1730s, that condensation forming on a surface layer of volcanic stones was generating sufficient water for the irrigation of vines. The stone mulch certainly promotes condensation by increasing the surface area on which dew can form, but also inhibits evaporation from the soil, perhaps by as much as 75%, although this is questioned by some commentators. Even if condensation is not capturing as much water as some people have suggested, mulching with stone is certainly a useful technique for growing in arid conditions.[28]

Air Wells

In the early part of the nineteenth century, I. Zibold, a forester working in Feodosia in the Crimea, experimented with dew catching. Zilbold had misidentified ancient Greek tombs as dew collection devices and had observed that condensation formed freely on the trees of the pine forests overlooking the Black Sea. Encouraged by these observations, he experimented with rubble-filled craters, which were the inspiration for high-mass air wells built by others during the twentieth century. A particularly attractive example of a high-mass air well is that built by Belgian engineer Achile Knapen in Trans-en-Provence in 1931.[29] The structure is 14m high and has perforated masonry walls 3m thick. There is a massive central concrete column. The concept was that warm moisture-laden air would enter the outer structure during the day and condense on the central concrete column at night. Unfortunately the huge structure produces just a few litres of water each night. Over the years, people have continued to experiment with similar structures; however, yields are routinely found to be too low to justify the cost of construction and the technique has therefore never gained widespread support.[30]

Lightweight Fog Catchers

More successful than earlier efforts to catch dew using massive structures have been some more recent efforts to capture fog using lightweight mesh. In 1992 experiments began in Chungungo in Chile, where annual precipitation is less than 60mm, to use polypropylene mesh to collect condensation for a reforestation project. The system mimics the way that leaves in the cloud forest also capture water. Early success with the experiment in Chungungo encouraged the erection of 94 mesh collectors as well as the provision of a storage tank and pipework to supply the 300 inhabitants of the village. The system produced an average of 15,000 litres of water each year for ten years. The project has since had a patchy history, having fallen into disrepair for a while because of the need to keep the nets under tension; however, the technique had been shown to work.[31] Encouraged by the pioneering work in Chungungo, the NGO FogQuest has set up similar operations elsewhere in Chile, in Yemen, Guatemala, Haiti and Nepal. Now there are projects in more than 25 countries.

Foil Collectors

The International Organization for Dew Utilization is promoting the use of foil or plastic dew collectors, which can be free standing or roof mounted. These collectors are lightweight and insulated and designed to have low mass so that they radiate heat into the sky at night. Rapid cooling promotes condensation and the collectors are usually tilted 30° to promote easy collection of water. There are similar installations, which have been configured to collect both dew and rain, with several projects in dry places around the globe, including Spain, Croatia, Israel, Tunisia, Morocco, Mauritania, India and Australia.[32]

Biomimicry: Desert Beetle

There is also the potential to use materials that increase the amount of condensate collected. The Namib Desert beetle harvests dew on its underside. This creature has alternating bands of microscopically rough and smooth surfaces on its exoskeleton, which creates alternating hydrophobic and hydrophilic regions, which are thought to be more efficient dew collectors than uniform materials. Chemical engineers at MIT have made coatings that mimic the banded hydrophilic and hydrophobic exoskeleton of the Namib Desert beetle, with the potential for several applications including dew catching.[33]

Potential in Towns

Although dew- and fog-catching techniques are aimed at reforestation projects or the provision of water in remote villages far away from conventional water supply networks, they also have potential for use in urban areas, perhaps creating arrangements where green roofs or living walls self-irrigate. In hot and dry urban locations where there is sufficient humidity in the air, it is possible to imagine nets or other filamentous structures catching and channelling water to plants. There may also be opportunities to use photovoltaic cells in such arrangements.

Condensate

Air conditioners and refrigerators pump refrigerants, which take heat from inside buildings or other enclosed spaces to the outside. As air passes across cooling coils, water condenses. This condensate is channelled away from the equipment and buildings so that it does not damage the structures or interiors. Condensate is usually piped away to drains, although dripping air conditioners are commonly encountered and often annoy downstairs neighbours in high-rise apartment buildings. A typical single window-mounted air conditioning unit will produce a meagre 5 litres of condensate a day, whereas a central air-conditioning unit servicing a single large dwelling might produce up to 75 litres a day. Air conditioning systems for larger buildings have the potential to produce even large quantities of condensate, around 40 litres per day for every $100\,m^2$ of office space. Condensate is very high quality water, free from minerals. It should not be used for drinking because of the contact with the equipment and the possibility of contamination, however it is suitable for use in irrigating ornamental plants (not plants intended for consumption).[34]

Individual homesteads in rural areas with sufficient rainfall can collect and store enough rainwater to be self-sufficient if there is enough surface area to collect the water and large enough tanks to store it. Although significant volumes of rainwater can be harvested in cities in a cost-effective way, in most climates urbanites will not be able to collect and store enough water to allow them to sever their ties with the water supply grid. Singapore's world-leading plans to meet 20% of demand through local rainwater collection illustrate that point. The recycling of urban water – the subject of the next chapter – has the potential to furnish much larger quantities.

12. Recycle

Huge Potential

There is huge potential to recycle water. Most wastewater enters the environment. In advanced countries wastewater is treated before it is discharged but there are very few places where wastewater is treated to a quality where it can be returned to reservoirs or direct to consumers. Singapore leads the world in the recycling of water. The city-state currently provides 30% of its water supply (about 270,000 m³ per day) from treated wastewater. Wastewater from conventional sewage plants, which would normally be discharged to watercourses or the sea is channelled to one of five plants (called NEWater plants locally)[1] where it goes through a three-stage treatment process, the stages being microfiltration, reverse osmosis and ultraviolet disinfection. Although the water is treated to international drinking water standards it is used primarily in industry and for the cooling of commercial buildings, although a small percentage is fed into reservoirs. The Singapore situation is extreme given its situation as a densely populated island without a hinterland however it does indicate what may eventually happen in many other places; that is, the treatment of wastewater to a very high standard and the separation of water supplies between domestic potable water and water used for commercial and industrial purposes.

The Water Sensitive City, First Edition. Gary Grant.
© 2016 John Wiley & Sons, Ltd. Published 2016 by John Wiley & Sons, Ltd.

Treated Wastewater

Even where wastewater is not treated to the standards of NEWater in Singapore, it is often used for irrigation purposes. Doha in Qatar, a country with no rivers and 80 mm of annual rainfall, which relies on expensive desalination plants for about half of its freshwater supply, has practised the reuse of wastewater since the 1950s. Treated wastewater makes up about 10% of the total supply and is used to irrigate the urban landscapes of Doha at no cost to the consumer. There are concerns over the inefficiency of this irrigation regime, with much of it wasteful surface irrigation, so the situation is far from sustainable; however, it seems likely that the use of treated wastewater will become an increasingly important component of the water management regime which will include more desalination, more efficiencies and the recharge of natural aquifers using treated effluent.[2]

The Big Dry

Australia suffered a severe drought in the early 2000s, known as the Big Dry.[3] Reservoirs in the Sydney area reached record lows and it was necessary to put curbs on some of the more wasteful uses of drinking water supplies. Golf courses, for example, were concerned that lush greens would become 'browns' and so the practice of sewer mining began. One example was the Pennant Hill Golf Club, which sought permission to tap into and extract water from a sewer, which serves about a thousand dwellings, which happened to flow under the golf course. Sewage extracted from the sewer is treated in a small treatment plant based on membrane bioreactor[4] technology. The treatment plant is hidden amongst trees on the course and the treated water is used for toilet irrigation and toilet flushing, thereby reducing the consumption of potable water on the course by more than 90%. Despite some initial scepticism and resistance, the success of that project has encouraged more than a dozen copycat schemes in the Sydney area.[5]

Greywater

Greywater, or sullage, is wastewater that emanates from sinks, showers and baths. It is relatively clean in comparison with sewage (occasionally referred to as blackwater). In conventional drainage systems, grey water is directed to sewers along with blackwater; however, it makes sense to separate and recycle grey water because it can be cleaned easily. Greywater is contaminated with soap, salts, phosphates and small amounts of organic matter including hair and skin, so it is clearly not fit to drink and should not be stored without prior treatment because of possible risks to health caused by decomposition of the organic content. There are, however, situations where greywater has been channelled directly onto gardens and plant nurseries, although this may be illegal in some jurisdictions as grey water may be classified as sewage.[6] There is also the risk of phosphates entering watercourses if the discharge point is close and if there is insufficient soil and vegetation to take up the phosphate before it reaches the aquatic environment. Phosphate is a notorious pollutant of aquatic ecosystems, causing algal blooms and loss of aquatic biodiversity.[7] Small volumes of relatively clean grey water entering large gardens in rural locations are unlikely to cause major problems, however. Where grey water is used directly in gardens, larger contaminants like coffee grinds and fats can be removed with simple straw traps.[8]

Treating Greywater

In most situations where it is recycled, greywater is treated before it is stored and reused in the house or place of work for toilet flushing or other nonpotable uses. Methods used to purify greywater include sand or lava filters, constructed wetlands, living walls, activated sludge systems and aerobic and anaerobic filters. Distillation, filtration using membranes and reverse osmosis are also used in larger systems. Treatment is usually undertaken in two or three stages with various techniques used in combination. The final treatment is usually sterilization by ultraviolet light. In a rural or suburban setting where the space is available constructed wetlands may be the most cost effective way of treating grey water, however in more urban settings where space is limited, membranes are the preferred technology.

A typical, modern, greywater system will include a pretreatment unit, which includes a coarse filter of beads, straw or wood shavings designed to remove fibre, hair and larger particles. Usually the filter material will be enclosed in a sack placed inside a metal containment vessel to facilitate the changing of the filter. A grease trap will also be necessary if wastewater is received from a kitchen sink. Greywater from kitchens contains food waste and this can become a problem if the waste accumulates, becomes anaerobic. Blockages and smells can become a problem without frequent maintenance. For this reason waste from kitchen sinks will often be omitted from greywater recycling systems. Whatever the situation, however, pretreatment units will require occasional cleaning, meaning that greywater harvesting is not a 'fit and forget' technology.

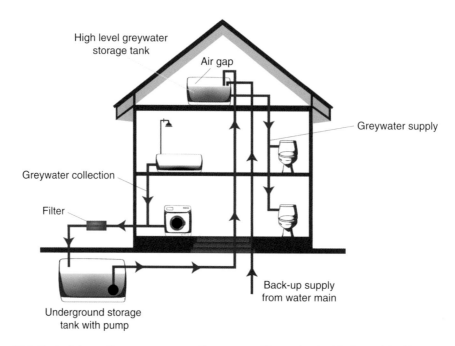

High level greywater
storage tank

Air gap

Greywater supply

Greywater collection

Filter

Back-up supply
from water main

Underground storage
tank with pump

Figure 12.1 Typical domestic greywater recycling system. Illustration by Marianna Magklara.

Microbes and Membranes

In a typical commercially available compact greywater treatment system, once coarse debris has been removed at the pretreatment stage, the grey water, still polluted with organic materials, will usually pass into an aerobic treatment chamber. The chamber contains a liquor of bacteria where air is pumped through to speed digestion of the organic waste. After spending some hours in an aerobic treatment chamber the wastewater can then be further cleaned by being passed through a membrane bioreactor.[9] Membrane bioreactors involve the use of membranes with very small pore sizes, typically 2500 times smaller than a human hair and small enough to filter out all bacteria and most viruses – in a process known as ultrafiltration.[10] The activity of microbes and the passing of bubbles across the surface of the membrane can keep it from clogging.

Regulations

Until recently, the use of greywater was not only extremely uncommon, but prohibited in some jurisdictions. Greywater has traditionally been considered a form of sewage because of the possibility of contamination with human pathogens; however, bearing in mind the relatively low content of human waste in greywater, the growing importance of water conservation and the availability of reliable technology to treat greywater, legislators in some jurisdictions have been persuaded to permit the practice and even promote it. While still not fully supported in the United Kingdom, there are states, including California, New Mexico, Utah and Wyoming, which now permit the use of greywater for subsurface irrigation. Since 2009, Montana has gone further than this and has allowed greywater recycling in businesses and multifamily dwellings, having previously (in 2007) only permitted the use of greywater in single-family homes.[11] In 2006 in Australia, the New South Wales government relaxed the relevant regulations to allow householders to divert greywater for irrigation without seeking the prior approval of the local authority, providing certain safeguards are followed. The advice now given is that approved equipment should be used, that untreated greywater should only be used for subsurface irrigation, that diversion devices be switched off when irrigation is not required (especially during rain) and that people should ensure that filters are cleaned regularly. The advice also warns against the use of untreated wastewater when someone in the home is sick and advises against the topping up of rainwater tanks or swimming pools. The advice also stresses that untreated greywater must not be allowed to enter surface drains.[12] The New South Wales government also requires that approval be sought from the local authority before a greywater treatment system is installed and that a qualified plumber must undertake the installation.

Standards

Since 2010, there has been a British Standard for greywater systems (BS 8525-1:2010).[13] The standard advises that grey water should only be used for toilet flushing, garden irrigation and washing machines and only then providing that it has been treated to an appropriate standard (which would usually be water of a quality that meets the requirements of the European Bathing Water Regulation 2006/7/EG).[14] Greywater treatment systems are still relatively new to the United Kingdom; however, there are now several specialist suppliers and contractors, with some of these bringing experience from Germany and other places where the technologies are more established. Where developers seek certification under the Code for Sustainable Homes (a British assessment scheme for sustainable buildings), greywater systems are increasingly being considered.[15] Where developers wish to install greywater systems in new developments, the water supply companies will review the drawings and specifications of each scheme and make an inspection upon installation. The main concern of the water supply companies is that there is separation between the supply of mains drinking water and any water recycling or rainwater harvesting scheme.

German Pioneers

The first officially approved greywater recycling scheme in Germany was undertaken in Berlin in 1989. Expert opinion was divided at the time, with some cautious support for the idea and others predicting a public health disaster caused by cross connections. In the early 1990s, the Technical University of Berlin (TU-Berlin) developed schemes to provide treated greywater for toilet flushing suitable for apartment buildings of 100 dwellings and by 1995 standards had been published. By 2005 an estimated 400 greywater harvesting systems were operating in Germany. Now there are thousands. Grey water systems in Germany are multistaged, with greywater entering a settlement tank before undergoing biological treatment and ultrafiltration, clearing (stilling) and finally ultraviolet disinfection. In Germany, popular alternatives to biological treatment in an enclosed vessel are constructed wetlands or vegetated beds, which occasionally feature as a component of the exterior landscape of a development. Although constructed wetlands have been shown to perform well, with water routinely treated to meet standards set by the European Union Bathing Water Directive, it should be noted that significant quantities of water do evaporate from planted beds, which reduces the volume of water available for subsequent reuse.[16] Another issue with external constructed wetland systems is that they may cease to operate in the very cold weather experienced at higher altitudes or closer to the poles. This problem has been overcome in New England by placing vegetated beds in greenhouses.[17]

Jordan

The Dead Sea Spa Hotel in Jordan is a four-star hotel in a water stressed location. Jordan is a popular destination for tourists; however, ground water levels in the River Jordan basin are falling. Neighbouring Syria, Lebanon, Israel and the Palestinian territories are all affected by chronic water shortages. When the hotel decided to expand in 2008, it was already relying on up to ten tankers a day to ensure that the water supply tank was replenished, so it was clear that something needed to be done to address the issue. Realizing that 80% of the wastewater generated by the hotel was greywater, the owners took the opportunity to commission a German designed and manufactured greywater recycling system. The new system, which cost $80,000, saves $5500 \, m^3$ of water a year and was the first of its kind in the Arab world. It is likely to become a project that inspires many others to adopt the technology.[18]

It makes economic sense to recycle greywater in larger complexes like hotels, offices and apartment blocks; however, the recycling of greywater at the domestic level has been questioned. The requirement to treat greywater to a sufficiently high quality to prevent insanitary conditions developing in the storage tanks means that costs tend to be too high for a typical household. With a minimum cost of equipment of £3000 and installation charge of £1000 and annual running costs of £150 (2014 prices), only the wealthiest homeowners can afford to install greywater harvesting systems.[19] However, cheaper systems, designed to produce water for toilet flushing only, are under development. When combined with systems that recover waste heat from shower and bath wastewater, it is predicted that these domestic-scale systems will begin to become attractive to consumers. Soon, it is hoped, systems that will cost £300 to buy and £30 a year to run will come onto the market.[20] Widespread greywater recycling at the domestic level will be particularly important for water conservation, because the technology has the potential to cut consumption by 30%.

13. Water Quality

Nature Cleans

Visit the wilderness and you expect water to be clean. It usually is. There may be localized temporary contamination where a dead animal has fallen into a watercourse or a more permanent but localized problem where birds roost or wild animals live communally, but the work of scavengers, soil fauna, bacteria, fungi and plants in soil and water means that potential pollutants of water are rapidly taken up into the cells of organisms. Most pollutants are nutrients and the rich assemblage of species that occur in natural habitats act in combination to consume, sequestrate and therefore clean watercourses – one of the essential regulating ecosystem services. Where people live in small numbers they can avoid contaminated water by moving around or by keeping supplies and waste separated, but as settlements and populations grow in number, water quality declines.

Safe to Drink?

The quality of water is traditionally characterized in terms of human needs. The proof that contaminated drinking water could lead to disease and premature death in the nineteenth century was followed by efforts to filter and disinfect drinking water in the twentieth century, but universally accepted standards for drinking water for the purposes of protecting public health are yet to be established, despite a

The Water Sensitive City, First Edition. Gary Grant.
© 2016 John Wiley & Sons, Ltd. Published 2016 by John Wiley & Sons, Ltd.

declaration in 2010 by the United Nations General Assembly that safe and clean drinking water and sanitation are a human right, essential for the full enjoyment of life and all other human rights.[1] The World Health Organization (WHO) has published guidelines, which include recommended maximum limits for contaminants including pathogens, naturally occurring chemicals, additives and radioactive materials, which are known to have an adverse impact on health.[2] The approach advocated by WHO is to manage the risk to human health, which could be caused by hazards that are known to affect the safety of drinking water. The WHO definition of safe drinking water is that water that does not represent a risk to health over a lifetime of consumption. There is also a recognition that infants and the elderly are at the greatest risk of being affected by waterborne disease. It is argued by the WHO that it would not be appropriate to adopt international standards because of the differences between cultures, the capacity of regulators and the differing priorities for the allocation of resources in different nations. The WHO recommends a holistic approach to safeguarding water quality, considering the catchment and source and prioritizing prevention of contamination over treatment (which is consistent with the ecosystem approach). Inevitably though, given the widespread contamination of water sources, treatment is necessary and the guidelines identify contaminants which should be removed and the maximum safe levels of those contaminants.

Microbes

The first category of contaminants considered under the WHO guidelines is microbes. The greatest risks from drinking water are associated with contamination by pathogens derived from faeces (both human and animal). The recommended strategy is to ensure that multiple barriers are established throughout the water-supply chain and that the quality of drinking water and water sources is monitored. In addition to pathogens carried in faeces (including bacteria, viruses and parasites) there are also other microbial hazards, which may occur in water, including nematodes, cyanobacteria and the Legionella bacterium. Pathogens in drinking water are contaminants of particular concern because of the diversity of characteristics and variations in virulence. There is also the possibility of multiplication occurring in people, their food and drink, or water systems in the home. Infectious diseases caused by pathogens are the most common and widespread risk to health caused by drinking water. The table overleaf lists those organisms for which there is evidence of transmission through water supplies. In addition to the organisms in the table there are more than a dozen other species where there is a suspicion, but not confirmation, that they might be transmitted in drinking water or that there have been isolated cases recorded. The number of pathogens transmitted by

Table 13.1 Pathogens transmitted in drinking water.

Pathogen	Severity of outbreaks	Persistence in water supplies	Resistance to chlorine	Infectivity
Bacteria				
Burkholderia pseudomallei	High	May multiply	Low	Low
Campylobacter jejuni; C.coli	High	Moderate	Low	Moderate
Escherechia coli (pathogenic)	High	Moderate	Low	Low
E.coli (enterohaemorrhagic)	High	Moderate	Low	High
Francisella tularensis	High	Long	Moderate	High
Legionella spp.	High	May multiply	Low	Moderate
Leptospira	High	Long	Low	High
Mycobacteria (non-tuberculous)	Low	May multiply	High	Low
Salmonella Typhi	High	Moderate	Low	Low
Other salmonellae	High	May multiply	Low	Low
Shigella spp.	High	Short	Low	High
Vibrio cholera	High	Variable	Low	Low
Viruses				
Adenoviruses	Moderate	Long	Moderate	High
Astroviruses	Moderate	Long	Moderate	High
Enteroviruses	High	Long	Moderate	High
Hepatitis A	High	Long	Moderate	High
Hepatitis E	High	Long	Moderate	High
Noroviruses	High	Long	Moderate	High
Rotaviruses	High	Long	Moderate	High
Sapoviruses	High	Long	Moderate	High
Protozoa				
Acanthamoeba spp.	High	May multiply	High	High
Cryptosporidium hominis/parvum	High	Long	High	High
Cyclospora cayetanensis	High	Long	High	High
Entamoeba histolytica	High	Moderate	High	High
Giardia intestinalis	High	Moderate	High	High
Naegleria fowleri	High	May multiply	Low	Moderate
Helminths				
Dracunculus medinensis	High	Moderate	Moderate	High
Schistosoma spp.	High	Short	Moderate	High

Source: After WHO.[3]

drinking water is likely to increase as previously unknown species continue to be discovered. In the last quarter of the twentieth century many new waterborne pathogens were discovered, particularly viruses, and other species, that had not been seen for many years, re-emerged.

There are fears that continuing population growth, the continuing pattern of dense urbanization and the effects of climate change may create the conditions for new pathogens to flourish.

Which Pathogens to Monitor?

With so many species of known pathogens it is not feasible to monitor and set targets for them all. In addition there are problems in detecting and measuring pathogens – in some cases a single organism in 10,000 litres of water may present a risk to health. Health authorities choose representative species to monitor, species that are believed to be reliable indicators of water quality. The selection of indicator pathogens varies from country to country and between climates. Rotaviruses, enteroviruses and noroviruses have all been selected as reference pathogens by various authorities. Rotaviruses are the most important causes of gastrointestinal infections in children and frequently cause death in developing countries. Rotaviruses are excreted in very large numbers by infected people; however, there are no routine culture-based methods for measuring the number of organisms in a standard sample. Enteroviruses can be the cause of severe diseases in children, including paralysis, encephalitis and meningitis. These are also excreted in large numbers from infected people and in this case there are culture-based methods for measuring the number of organisms in a standard sample. Noroviruses cause acute but short-lived gastroenteritis in all age groups. There are no culture-based ways of measuring the abundance of noroviruses in water but there are models, which have been developed for the purpose of estimating infectivity.

Bacteria

The bacteria are the group of pathogens most easily killed by disinfection. In addition most bacteria do not survive for long outside of the gut. *Vibrio cholera* is one species that may need to be monitored. It can cause diarrhoea and often proliferates in the conditions that follow war and natural disasters. Although the dose required for infection is high, people often die following infection. *Campylobacter* is a common organism in the wider environment and a widespread cause of diarrhoea, however fatality rates are relatively low. Waterborne transmission of *Escherichia coli* (usually shortened to *E. coli* and notably strain O157, however there are other strains) is uncommon but infection can be fatal. A small dose (fewer than 100 organisms) can lead to infection. In developing countries more than 2 million people are infected by *Shigella* each year and of these, about 60,000 go on to die.

Shigella is another organism that can trigger an infection with fewer than 100 organisms. *Salmonella typi* is a particular species of bacterium, which has been the cause of large outbreaks of typhoid.

Protozoa

Protozoa are the most resistant group to chemical disinfection. Some species can be controlled by ultraviolet light. The cells are relatively large though ($>2\,\mu m$) and can be removed by filtration. Unlike most pathogenic bacteria and viruses, protozoans can survive for long periods in water and infective doses are usually low. Genera of protozoa frequently associated with cases of diarrhoea are *Giardia* and *Cryptosporidium*.

Treatment

With the exception of supplies of very high quality waters from aquifers, treatment is required for drinking water supply in order to remove or kill microbes. Normally several stages of treatment are required. The table overleaf sets out the various treatment processes, which are commonly used.

As well as the treatment processes included in the table there are a number of other methods used at the smaller domestic scale. As well as good treatment, it is also important that, where people do receive piped drinking water, that it is stored carefully in their homes. The most common form of household treatment is chlorination. This is usually achieved by using chlorine in the form of hypochlorous acid (household bleach) or a dilute sodium hypochlorite solution. Free chlorine is usually applied in a household situation at the rate of 2 mg/l for clear water and up to 4 mg/l for turbid water. Occasionally iodine, another strong oxidant, is used, although care has to be taken with long-term use because of the uptake of iodine by the thyroid gland. Filters (including ceramics and membranes) with known pore sizes are also commonly used. The use of granular filters (for example those filled with coarse sand) is another popular technique occasionally found in domestic situations. Sand works through a combination of filtration, biological action and physical and chemical processes which clean the water. A biologically active film (known as the schmutz-decke)[4] retains pathogens and causes their degradation. In common with large-scale treatment facilities, ultraviolet lamps (in particular those that produce light with the most germicidal wavelength of 254 nanometres) are also used. Other less common approaches to disinfection in the household setting include solar disinfection, where sunlight is used. This usually works by passing light through the water

Table 13.2 Treatment techniques.

Process	Pathogen	Remarks
Pre-treatment		
Roughing filters	Bacteria	A series of graded gravel beds
Storage reservoirs	Bacteria Protozoa	Settlement of sediments, sunlight breaks down some material
Bank filtration	Viruses Bacteria Protozoa	Pathogens filtered by vegetation, filtered and broken down in soil
Coagulation, flocculation, sedimentation		
Conventional clarification	Viruses Bacteria Protozoa	Iron and aluminium coagulants added to water in basins
High-rate clarification	Protozoa	Polymer used to speed flocculation and save space
Dissolved air flotation	Protozoa	Bubbles take material to surface to be skimmed off
Lime softening	Viruses Bacteria Protozoa	Calcium hydroxide precipitates calcium and magnesium ions – promotes flocculation
Filtration		
Granular high-rate filtration	Viruses Bacteria Protozoa	Sand, gravel or other coarse material holds back suspended solids and microbes are decomposed
Slow sand filtration	Viruses Bacteria Protozoa	Can remove more pathogens than coarse material
Pre-coat filtration	Viruses Bacteria Protozoa	Solids are cut from a rotating drum pre-coated with diatomaceous earth or perlite. Water drains through drum. Relatively free from clogging.
Various membranes	Viruses Bacteria Protozoa	Pores in membranes allow water to pass through but hold back pathogens.
Disinfection		
Chlorine	Viruses Bacteria Protozoa	Not effective in turbid water. Not effective against Cryptosporidium oocysts
Chlorine dioxide	Viruses Bacteria Protozoa	Effective against Giardia and Cryptosporidium
Ozone	Viruses Bacteria Protozoa	Viruses more resistant than bacteria
Ultra-violet light	Viruses Bacteria Protozoa	Intensity must be adequate and wavelength correct. Not effective in turbid water.

but also by heating the water at the same time. Heating is a well established means of sterilization. A rolling boil for at least a minute kills germs and protozoa but uses a lot of energy and is only really practical for treating small volumes. It is also important to note that boiling will not rid water of chemical contaminants.[5]

Chemical Contaminants

It is normal for drinking water to contain a range of organic and inorganic chemicals in low concentrations. At naturally occurring levels these chemicals are not believed to affect human health, although some can affect taste. Examples of chemicals that affect taste include bromide, chloride, hydrogen sulphide, iron, manganese, molybdenum, potassium, sodium and sulphates. Naturally occurring chemicals which present a cause for concern, include arsenic, barium, boron, chromium, fluoride, selenium, and uranium. Arsenic is a special case because safe levels are particularly low (0.01 mg/l). In some countries, including Bangladesh, contamination of well water with arsenic has been recognized as a public health emergency. Bangladesh has seen the largest poisoning of a population in history with millions of people exposed to arsenic in drinking water.[6] There are now many chemicals used in industry, households and in agriculture, which can become contaminants in drinking water. In addition, many chemicals used to treat water (notably chlorine) can react with other naturally occurring chemicals, pipework or contaminants. In some regions pesticides may be deliberately added to water (for example mosquito larvicides). There are too many different kinds of chemicals to make widespread monitoring feasible. Fortunately serious contamination is rare and when it does occur, it is usually diluted sufficiently to be below concentrations where it is likely to cause harm. There are, however, a number of chemicals that do frequently occur and which are known to pose a threat to health and for which guidelines have been established.

Nitrates

Nitrates in watercourses and lakes can be elevated because of agricultural runoff. As well as damaging aquatic ecosystems, nitrates can enter drinking water supplies and become a threat to human health. High levels of nitrate in drinking water can cause infants to become seriously ill or die. Methaemoglobinaemia (or blue-baby syndrome as it is commonly known) occurs when nitrate is converted to nitrite in the gut and subsequently interferes with the absorption of oxygen by the blood. In the United States, drinking water must contain no more than 10 mg of nitrate per litre[7] although the WHO and European Union standard is 50 mg of nitrate per litre.[8]

159

Table 13.3 Chemicals of health significance in drinking water.

Chemical	WHO Guideline	Notes
Inorganic	mg/litre	
Boron	0.5	Guidance level provisional. Occurs naturally. Byproduct of glass, soap and detergent manufacture. Flame retardant. Affects male reproductive tract.
Copper	2	Guidance level provisional. Used in electrical and plumbing systems. An essential nutrient. Can cause acute gastrointestinal problems in high doses.
Nickel	0.02	Guidance level provisional. Used in steel alloys. Metallic nickel possibly carcinogenic. Intake more often from food than water.
Nitrate (NO_3^-)	50	A naturally occurring chemical and fertiliser. Toxic when reduced to nitrite.
Nitrite (NO_2^-)	3 (acute) 0.2 (chronic)	Can cause cyanosis and asphyxia in babies.
Uranium	0.002	Guidance level provisional and relates to chemical not radiological effects. Intake more often from food than water. Potentially carcinogenic.
Organic	µg/litre	
Benzopyrene	0.7	Usually enters the environment through combustion but has entered drinking water from coal-tar applied to distribution pipes. A carcinogen.
Edetic acid (EDTA)	600	Used as a food additive, in medicines and personal care products. Toxic in high doses.
Microcystin - LR	1	Provisional guidance level. A cyanobacterial toxin, which affects the liver.
Pesticide	µg/litre	
Bentazone	300	A herbicide, which is only toxic to humans in very high doses.
Carbofuran	7	An insecticide, which is a potentially fatal neurotoxin in humans.
Cyanazine	0.6	A persistent herbicide, which causes genetic mutations and birth defects and is a carcinogen.
1,2-dibromoethane	0.4–15	Provisional guideline figure. An insecticide. Formerly added to petrol as an antiknock agent. Causes birth defects and extremely carcinogenic.
2,4-dichlorophenoxyacetic acid (2,4-D)	30	A commonly used herbicide, which is a potential carcinogen.
1,2-dichloropropane (1,2-DCP)	40	An insecticide and solvent. No longer widely used. A potential carcinogen.
Diquat	10	A herbicide which is toxic if ingested in large quantities.
Pentachlorophenol	9	A fungicide and potential carcinogen.
Terbuthylazine (TBA)	7	A herbicide which may be toxic if ingested in large quantities.
Disinfectant by-product	µg/litre	
Chloroform	200	Formed during chlorination. Can cause liver damage and is a potential carcinogen.

Pharmaceutical Contaminants

An emerging issue in drinking water quality is that of contamination by pharmaceuticals (including antibiotics, hormones and antiepileptic and antianxiety medications). These chemicals enter the environment through disposal into sewers, by being passed in faeces or urine or can be part of agricultural runoff (large-scale use of pharmaceuticals is a feature of modern industrial farming). Treatment of drinking water does not remove all of these chemicals and targeted investigations of particular pharmaceuticals have shown that municipal drinking water is usually contaminated.[9] Systematic monitoring of pharmaceuticals does not occur, so the problem is yet to be fully described and understood. The official view is that concentrations are too low (usually less than 0.1 µg/l) to be a public health concern. At this time, the most worrying effects to be described in detail are the effects of the active ingredients of birth-control pills on the sexual development of freshwater fish and amphibians.[10]

Radioactive Substances

If drinking water contains radioactive substances it can constitute a threat to human health. The risks, however, are small in comparison with those presented by microbes. Radioactive substances may be naturally occurring or man made. In practice more can be done to manage the risk presented by man-made radioactive substances by containing them and excluding them from the water supply in the first instance. Little can be done to exclude naturally occurring radioactive substances from aquifers and avoidance of water from those formations may be necessary. Naturally occurring radionuclides include potassium-40 and the products of thorium and uranium decay. Radon can be released from water and ingested as a gas. Naturally occurring radon in certain regions accumulating in the air inside buildings can cause lung cancer and is more of a concern than radon in drinking water.[11] Where water supplies are radioactive, clearly they should be avoided; however, it is possible to filter up to 100% of radioactive substances from drinking water by using various treatments including coagulation, sand filtration, ion exchange and reverse osmosis.

Smell and Taste

The smell, taste and general appearance of drinking water can be an indication of problems but these may not be reliable indicators of quality. Some discoloured or unpleasant tasting waters may be safe to drink, however suppliers try to avoid creating situations where people turn away from safe but tainted water in favour of other sources of water that is not safe to drink. There are difficulties, however, in establishing guidelines for the colour, taste and smell of drinking water because of natural differences to which people may be accustomed.

Standards

In most parts of the world there may be guidance on drinking-water quality; however, apart from a few exceptions, there is no legislation to require that drinking water is safe. Those exceptions include the United States, the European Union and China.

United States

The United States Congress passed the Safe Drinking Water Act in 1974. There had been earlier legislation, for example when, in 1914, the federal government set limits for the bacteriological quality of drinking water on ships and trains. Later initiatives during the twentieth century expanded on these regulations but it wasn't until 1974 that all public water supplies were covered. Surveys made of public water supplies in 1969 indicated that only 60% of municipal water supplies were meeting public health standards and so the drafting of legislation began.[12] Originally the emphasis for the regulators was to require adequate treatments to ensure that drinking water is free from pathogens, however later amendments, especially those that were passed in 1996, recognized the importance of protecting water sources. This amounted to an acknowledgement that treatment alone cannot be relied upon as a cost-effective approach to the task of providing safe drinking water. The 1974 legislation set enforceable maximum limits for contaminants, including microbes, the byproducts of disinfection, metals, nitrate and poisons like arsenic.[13]

Europe

The European Drinking Water Directive of 1998 set legally binding standards for the quality of water destined for human consumption in that continent, including water supplied to taps through municipal distribution networks, water in tankers, bottled water or water supplied for use in the food-processing industry. The legislation identified a total of 48 microbiological and chemical parameters that must be tested for. The parameters were taken from the World Health Organization's guidelines and the European Commission's own Scientific Advisory Committee. As is often the case in Europe, Member

States may be excused from meeting certain quality standards for a time (in a process known as a derogation) providing there is no threat to human health and providing water quality of the affected area cannot be maintained by any other reasonable means.[14]

China

China passed legislation requiring certain national drinking water quality standards in 2007. The standards set were based on World Health Organization guidance, which meant that actual water-quality levels were not up to standard in most cities. This led to a delay in adoption of the new standards until 2012, after the government had budgeted RMB 700 billion (US$112 billion) for upgrades to drinking water treatment and distribution infrastructure. The government had announced that it expected standards to be met by 2015, although this is still yet to happen. Although most major cities on the east coast of China have safe water supplies, there are some places with problems, in particular smaller cities and rural areas.[15]

Clean Water Act

Part of the difficulty for municipal water supply companies in meeting standards is the poor quality of water abstracted from the wider environment. The Federal Water Pollution Control Act, also known as the Clean Water Act, passed by the United States Congress in 1948 and amended in 1972, is an example of an effort to maintain and restore the quality of water in the wider terrestrial environment. By 1945, the Surgeon General had become concerned about the poor quality of water in many of the nation's rivers and the affect that this could have on health. The Senate Committee on Public Works had reported with understandable alarm that 3500 communities across the United States were disposing of 2.5 billion tons of raw sewage into watercourses. In addition, waterways had become dumping grounds for solid waste, with the only legislation controlling this (the Rivers and Harbors Appropriations Act of 1899), being concerned with maintaining navigation routes. The Senate Committee on Public Works declared, 'pollution of our water resources by domestic and industrial wastes has become an increasingly serious problem due to the rapid growth of our cities and industries ... polluted waters menace the public health through contamination of water and food supplies, destroy fish and game life, and rob us of other benefits of our natural resources.' The legislation was not universally welcomed and was weak, with no standards and no enforcement. At least the new legislation had put the issue of the quality of water in the wider environment on the agenda.[16]

When an oil slick on the biologically dead Cuyahoga River in downtown Cleveland, Ohio, caught fire in 1969 and damaged two rail bridges, it was not the first time this had occurred. The river had first caught fire in 1868 and there was also a more serious fire than the 1969 event in 1952, but the resulting publicity in the national press, including *Time* magazine, in 1969, embarrassed some legislators and provoked action at last. The National Environmental Protection Act, which established the Environmental Protection Agency, was signed into law in 1970. Intense activity continued and the Federal Water Pollution Control Act was effectively rewritten in 1972. During the ensuing years, around $3.5 billion was invested by the Northeast Ohio Regional Sewer District in new sewers and other clean up initiatives. Although there is still more to do, the river now supports more than 60 species of fish.[17]

Water Framework Directive

The European Union adopted the Water Framework Directive in 2000.[18] The objective was for member states to achieve 'good (chemical and ecological) status' for all groundwater, water bodies and coastal waters by 2015. This marked a departure from the earlier approach of setting maximum limits for pollutants, with a new emphasis on better management of river basins. The legislation requires that plans be prepared that not only prevent deterioration by reducing discharges of pollutants but must also bring about restoration of water bodies. There is also a new emphasis on protecting ground waters and addressing the problem of nonpoint sources of pollution (that is pollution from many sources over a wide area). The most commonly cited example of nonpoint pollution is agricultural runoff. Other examples of nonpoint source pollution include air borne pollutants, which may be brought to earth by rain and surface-water runoff from roofs and streets.

The status of surface waters is assessed under the Water Framework Directive by description and measurement of a range of parameters and features including:

- biological quality (by surveying aquatic flora, aquatic invertebrates and fish);
- geomorphological conditions (including river banks, river bed and continuity);
- physical qualities (including temperature and oxygenation);
- chemical qualities (including maximum allowable concentrations of a total of 41 different pollutants).[19]

The Water Framework Directive classification scheme for surface waters, which uses the parameters described above, has five classification categories, namely: high, good, moderate, poor and bad. The target is for all surface waters to be good.

Earlier Legislation

The Water Framework Directive was preceded by earlier legislation, beginning with the Surface Water Directive in 1975 and followed by the Drinking Water Directive in 1980. The latter legislation set binding water quality targets and the emphasis in that early wave of European legislation on water was on public health. A later wave of legislation, during the 1980s and 1990s, changed the emphasis towards pollution control and a common environmental policy for Europe. An example of this was the Urban Wastewater Treatment Directive of 1991 and the Nitrates Directive of the same year. These initiatives, combined with an update of the Drinking Water Directive in 1998, stimulated substantial investment in water treatment infrastructure, however there were some problems, with some member states delaying the translation of the European directives into national legislation. In some cases there were low levels of compliance, most notoriously with the nitrates entering watercourses, despite the imposition of heavy fines on member states by the European Union.[20]

The Struggle for Compliance

The adoption of the Water Framework Directive, then, was an opportunity to rationalize earlier disjointed regulation but also marked a belated recognition by policy makers that water knows no political or bureaucratic boundaries. Not surprisingly, implementation has been difficult. Nations tend to claim the right to use water within their borders and public awareness of European legislation is limited. There is also the problem of low compliance with standards and quality targets. The situation has been further complicated by the growth of privatized utility companies, which tend to prioritize their own economic interests over the greater good. There are also concerns over the high cost of creating regulatory regimes, something that the European Union does support through grants, but the full costs are not met, leaving national governments to fill the funding gap.[21] A holistic approach to water management has arrived in Europe; however, not everyone is participating.

Nonpoint Source Pollution

Water that enters surface water drains in urban areas is the main component of nonpoint source (NPS) pollution. There are difficulties with identifying the specific source of these pollutants, which include soils, atmospheric dust that has settled on streets and buildings, dusty from construction sites, dust from vehicular traffic, fuel and lubricants, pesticides, fertilizers and faeces from pets and urban wildlife. These materials tend to be washed into drains and from there into watercourses when rain falls after a dry period – the so-called first flush. Nonpoint source pollution is a major contributory cause of poor water quality in urban watercourses. Even in cities, much of the dusty material that collects in streets is derived from soil, including landscaped areas, planters, tree pits or excavations made to repair underground services. Most of this material is relatively benign, however the sediment that it forms can block drains and damage aquatic ecosystems by smothering vegetation and making water too turbid for some species.

Dust in the Streets

Part of the dust in streets, however, is caused by motor vehicles and includes worn-down road materials like asphalt, worn tyres, dust from brake pads, soot from exhausts and drips and spills from engines and fuel tanks. Dust created by motor vehicles includes iron, zinc, lead, calcium, chlorine, bromine, chromium, potassium, sodium, nickel, thallium, sulphur and carbon amongst others.[22] Lead is toxic and is therefore of particular concern. The amount of lead in street dust was significantly reduced following the removal of lead from petrol in developed countries in the 1980s and 1990s, in Africa and Indonesia in 2006 and finally in Algeria in 2015.[23,24]

Urban Runoff

Streets and parking areas are the main sources of pollutants that enter watercourses, increasing suspended solids in streams and pushing up biological oxygen demand (BOD), although landscaped areas can contribute significant amounts of phosphate, particularly if nutrient-rich soils have been used. In suburban areas, particularly in American cities, about half of the phosphates found in streams have been found to emanate from lawns, and this has led to the imposition of restrictions on the use of lawn fertilizers in some cities.[25] Roofs, especially those on older buildings that use copper, zinc and lead waterproofing, flashings and downpipes, can be the source of pollution, exacerbated by acid rain, which corrodes the metal. Most cities make use of herbicides (notably glyphosate) to control weeds on pavements and railways, with reports of herbicides even being used on roofs in some cities.

A Continuing Problem

Analysis of urban nonpoint source pollution around the world has shown that motor vehicles, particularly those that are powered by internal combustion engines, are a major cause of the problem. The switch to electric-powered vehicles (now under way) will reduce the problem but tyre and brake wear will continue to cause problems, especially with the release of zinc. Slower speeds, more cycling and walking and use of public transport will also lead to further improvements. Despite improvements to transportation systems and wiser selection of building materials, there will always be some atmospheric dust, soils and pollutants washing down downpipes and across paved areas. That is why it is so important that features are incorporated into surface water drainage systems to intercept and decompose pollutants before they enter watercourses (as described in Chapter 9).

14. Future Water-Sensitive Cities

Waste Not

In order to create future water-sensitive cities, per capita consumption of water of drinking quality will need to fall. Expectations of an ever greater consumption of water will end and those who continue to assume that supplies can be readily increased without negative consequences will be in the minority. The wasteful use of potable water in irrigating ornamental landscapes or washing private motor vehicles, for example, will be forbidden or will become socially unacceptable. Overall, more water will be saved through a gradual change in attitude. There will be changes within the home. For example, people will use water-efficient fittings, including shower heads and taps. Clever design will mean that most people will not notice any difference between the new efficient fittings and the old ones that they have replaced. There will also be more use of gadgets that alert people when they have used too much water, for example, when they have spent too much time in the shower or have put too much water onto the garden. Washing machines and dishwashers will be much more water efficient that those now in use. There will be no noticeable difference in the cleanliness of clothes or crockery.

The Water Sensitive City, First Edition. Gary Grant.
© 2016 John Wiley & Sons, Ltd. Published 2016 by John Wiley & Sons, Ltd.

Measure

Metering will be universal and water consumption data will be monitored in real time, enabling citizens, city authorities and suppliers to identify potential problems at an early stage. Municipal water supply networks will be better monitored and maintained as part of the Smart City approach, so that losses through the distribution system will be reduced to less than 10% – much lower than the 25% still experienced in British cities, for example (a figure that has not improved for the last 30 years).[1] It will be possible and necessary to bring down daily per capita household water use from more than 500 litres (typical of people in the United States and Australia) to levels below 100 litres (consumption levels that have already been achieved in the Baltic States and some German cities) or perhaps even the 80 litres of some Chinese cities.

Water Collection

Citizens of the future water-sensitive city will collect some of the water that they need from the roofs of their homes and the roofs of their places of work. Some of the water from plazas and streets will be deliberately channelled into tanks and water bodies where it can be stored and subsequently reused for irrigation of parks, gardens and toilet flushing. Installation of water butts to store rain from downpipes is already popular and will become increasingly common. It is also now commonplace for new commercial and residential developments to include large-scale rainwater harvesting systems. The larger rainwater harvesting systems of the kind being encouraged through schemes like the United Kingdom's Code for Sustainable Homes (CfSH)[2] have the potential to reduce mains water consumption by 50%. In the water-sensitive city of the future, adherence to standards for water conservation like those described in the CfSH will be commonplace.

Recycling and Cooling

In the typical apartment block, once the waste water disappears down the drain from the shower or sink, it will enter a greywater treatment facility in the basement. That water will be cleaned and recycled for toilet flushing and also to irrigate the planting on the outside of the block, for example, a living wall. The average resident will not give the treatment of greywater much thought, except, perhaps, when he gets a message about some interesting water management event that has occurred. An example might be how, during a heat wave, a resident receives a message on his smartphone, sent by the building management system, about how the computer controlled irrigation system has responded to a heat wave by delivering extra water to the apartment's living wall in order to provide extra evaporative cooling.

Smart Plumbing

Buildings themselves will include more pipes – pipes to channel rainwater from roofs to lower level cisterns, pipes to channel greywater to treatment units and pipes to irrigate gardens, interior planting and facades. There will also be software that monitors and controls water use, warning of leaks. Residents will be aware of how much water has been used because of smart metering and management systems that will keep you in touch with personal and collective water consumption rates.

Water and Power

The water-sensitive city will also be a low-carbon city. More of the electricity that powers the lights and appliances will be generated by photovoltaic cells or other renewable means, processes that use no water during operation, unlike the electricity from the fossil-fuel powered thermal power stations, which require water to produce steam and for cooling.

Figure 14.1 Biosolar roof in Switzerland. Photograph by Dusty Gedge.

Water and Roofs

Changing attitudes about the relationship between water, nature and buildings are beginning to change the appearance of buildings and this process will continue in the water-sensitive city. Steeply pitched roofs make no sense in terms of reducing and slowing the flow of rainwater. Green roofs are easier to establish on flat or gently sloping roofs and flat green roofs make better use of space for people and wildlife. A typical extensive green roof intercepts about half of all the rain that falls upon it in a typical year, with rain stored in the soil and subsequently evaporating or being transpired through vegetation. This provides useful cooling and water that does leave green roofs can go on to be collected and stored. When viewed from the air the water-sensitive city of the future will be a more verdant landscape than that we are used to. There will be more rooftop vegetation, including green roofs combined with photovoltaic panels - the so-called biosolar roofs[3] as well as drought-tolerant native vegetation at ground level. An attempt at passing a new law in France in 2015, which would have required all new commercial buildings to have either green roofs or rooftop solar panels has really brought it home to urban designers, architects and the wider public that rooftops are changing for good.[4] Looking out of windows in high spots above the typical city, people will be able to see the dozens of extensive green roofs, providing stepping stones for once rare species like certain butterflies, which will have become a feature of the roofscapes.

Figure 14.2 Biodiversity will be an essential part of the water-sensitive city. Photograph of common blue butterfly by Dusty Gedge.

Figure 14.3 High-rise greenery in Singapore: The Khoo Teck Puat Hospital. Photograph by the author.

Water and Walls

Eventually water leaves the roofs and there are further opportunities to green the water-sensitive city by using terraces and facades. Water can be directed from roofs straight to the roots of plants on facades or it can be harvested, stored and sent to plants through carefully regulated irrigation systems. People like to see out from buildings and natural light will continue to be in demand; however, spaces between windows and shade structures on the facades will be vegetated on most of the new buildings of the future water-sensitive city. These green facades will help to increase the area of vegetation within the city, having a profound positive affect on temperatures, air quality, wildlife and mental health. The treatment and greening of facades will become even more important as the number of tall buildings continues to increase. The planners, architects and horticulturalists of Singapore, the city that leads the world in the creation of high-rise greenery, are showing us what is possible.

Blue-Green Infrastructure

Integration between the green infrastructure network of greenspace and the conventional infrastructure network of pipes will improve. The green infrastructure network will become the blue-green infrastructure network, with better integration of waterways and drainage systems with green space. Apart from relatively small pockets where food is grown or fields are required for sports, vegetation in the wider terrestrial environment will not be heavily irrigated. An increasing appreciation and understanding of the ecosystem approach means that there will be movement towards the establishment of more areas of semi-natural vegetation, where native species growing in natural associations will be better suited to the climate than imported ornamental species. These seminatural greenspaces will be comprised of species that provide shade and beauty as well as habitats for wildlife. They will require less maintenance than purely ornamental landscapes. The groves and flower gardens that replace lawns will consume less energy because mowing will no longer be required. Most green spaces and street tree pits will be reconfigured as rain gardens in order to receive surface water runoff when it does rain heavily. Soils will be amended so that more water will be stored in the soil. As well as changing and improving green space vegetation, these features will reduce flood risk and people will be aware of this.

Figure 14.4 Biodiverse extensive green roofs are established on all new, large buildings in Basel, Switzerland. Photograph by the author.

Making Room

Where there is insufficient space for trees or tree roots in narrow streets or under pavements choked with pipes and cables, there may still be room for small rain gardens or planters fed by downpipes. Combined with the efforts to rid streets of unnecessary signs and street furniture, the new consensus that street-level vegetation is beneficial and desirable will inspire people to seek to transform unused and underused space.

A More Permeable City

Restored streams and rivers will thread their way through cities, connecting open spaces and connecting neighbourhoods with footpaths and cycle ways, making the city more permeable to wildlife and people. The reduction of consumption in cities will reduce pressures on the wider countryside creating the possibility of urban green infrastructure networks joining with regional and national green infrastructure networks. Cities will work closely with other communities within shared river basins.

Green Streets

The streets of the future water-sensitive city will be somewhat greener than before. There will be more trees and the rain gardens will mean that more vegetation will be seen. There will be channels and rills in the paving to channel water into rain gardens and tree pits. Where downpipes once ran straight into the ground, they will have been intercepted, running through planters and tanks set within the curtilage of buildings. With new buildings, downpipes will not usually be evident, with rainwater directed into cisterns located within or below the structure.

Street Life

During the twentieth century, precedence was given to the needs of people using motor vehicles. The offence of jaywalking was created. It has been suggested by Peter Norton that in the United States, until 1923, newspaper articles tended to blame motorists for accidents and yet by late 1924 the culture has changed, with jaywalkers more likely to take the blame.[5] Streets were sealed, smoothed and widened and priority was taken away from the pedestrian and given to the motorist. As streets were improved for vehicles, new surface water drainage systems were installed, drying the streets and polluting watercourses.

Figure 14.5 De-paving and greening the street in the Bankside district, London. Photograph by the author.

In many cities the increasingly demanding requirements of motorists and road planners led to the loss of verges and trees. The process of sealing has continued, even spreading from the urban core to the suburbs, where front yards have been turned into parking spaces – a process sometimes described as urban creep.[6] However the rebalancing of streetscapes in favour of the comfort and wellbeing of the pedestrian is already underway. Work by Jan Gehl and others has led to the creation of calmer, less cluttered streets where motorists must give priority to the pedestrian.[7] This new emphasis has provided more opportunities for people to find space for more trees and ground-level vegetation. Where streets continue to remain busy with traffic, the importance of trees and other forms of vegetation is being recognized again in many cities, where the last great large-scale planting efforts took place before motor vehicles became popular. Since the 1990s, with the establishment of organizations like Trees for Cities, there has been a concerted effort to persuade urban authorities to plant more trees.[8] Now, inspired by the pioneering work of Bjorn Embren and others in cities like Stockholm, more thought is being given to the possibilities of helping trees to survive in the urban realm by channelling water into carefully designed and enlarged tree pits or trenches rather than straight into the drains.[9]

Sparkling Streets

The clever use of water in streets goes beyond tree pits and rain gardens. When water is clean and sufficiently abundant, it can be run through pavements or roadside channels, creating local distinctiveness, providing habitat and drinking opportunities for birds and other wildlife. Water can enchant the passer by. It was once commonplace for streams to be run along streets in hillside towns for the convenience of inhabitants, who may not have enjoyed the convenience of piped water. As awareness of waterborne disease became widespread and piped supplies of drinking water arrived, these surface water features tended to be diverted underground, but they survive in some places, for example along the main street in Chard in Somerset, England. In the water-sensitive city, lost channels will be restored and new features created.

Urban Food Revolution

Many residents of the future water-sensitive city will enjoy looking after their own vegetable plots, being able to grow and swap and enjoy a wide variety of salads, herbs, vegetables and fruits. Previously neglected corners of the city, including, for example, roofs, window boxes and even walls are being used to grow fruit, vegetables and herbs. This will improve the diet of many people but will also reduce their water footprint, reducing the amount of virtual water consumed when foodstuffs are imported into the city. The connection between water and food will be evident at the average vegetable garden, where irrigation will be provided from water butts, filled with water drawn off from downpipes.

Urban Farms

Most initiatives are in the household and in the neighbourhood, however there are also larger scale urban farms in large rooftop greenhouses, like the Lufa Farms founded in 2009 in Montreal, Canada.[10] The benefits of urban agriculture include improved social cohesion, health and wellbeing, reduced transportation and energy costs and improved air quality. These are all good reasons for encouraging the practice; however, urban farming will also be able to use harvested rainwater and treated greywater, helping to accelerate the adoption of these techniques.

Agricultural Reform

Across the globe, agriculture is responsible for the greatest demand for water (70% of water abstracted from the natural environment is used in agriculture).[11] Citizens of the water-sensitive city will help to reduce this by adjusting their diet. They will eat more fruit and vegetables and less meat as they learn that the production of 1 kg of beef requires 15,500 litres of water, in contrast with the less than 1000 litres that is required to produce 1 kg of most types of fruit and vegetable.[12] Interest in water-wise healthy eating will be further encouraged as knowledge spreads of the ecological impact of heavy meat consumption, the cruelty suffered by many animals raised for meat and the increased risk of gut cancers[13] and heart disease.[14] Healthier diets and the shift from meat to vegetables will have an impact on farming and water management in the hinterlands of cities.

Relax and Play

Clean water is not to be feared and bringing clean water into the streets can entertain and promote play, bring joyful crowds and boost business. Clean water can be stored, channelled and pumped to fountains and pools when required, for example during hot weather or when special occasions or festivals create the demand. Some features may be turned on at certain times, for example, to allow school children to play with water when they come home from school in the afternoon. Play areas will become increasingly sophisticated, with water and vegetation featuring as well as the more conventional play equipment that tends to be set within relatively barren settings in contemporary parks and gardens. The traditional fountain where water complements sculpture as a spectacle will have been supplanted by modern computer-controlled installations where multiple jets encourage children to dash, dodge and splash. The future water-sensitive city will put more people in touch with water, meaning that they will be able to actually touch water.

Swimming and Boating

Moving away from the streets and piazzas, to the parks and waterways, there are waterbodies deep enough for people to be able to launch boats. Water quality improvements, which began in the 1970s in developed countries, have continued. Water is now clean enough for people to be able to safely launch kayaks and more and more cities are providing launching sites for kayaks and other boats.[15] [16] For those who prefer to relax by the shore close to home, more and more inland and urban beaches (as pioneered in Paris in 2002) will be established, providing places for adults to relax and children to play during the summer holidays.[17] Once further improvements in water quality have occurred, people can venture into the water to swim. If water quality has yet to meet the required bathing water quality standards, a separate pool can be provided (as it is at Paris-Plage), close to the water's edge or perhaps floating on the river itself (like the Badeschiff on the River Spree in Berlin).[18]

Encounters with Nature

Some prefer the quiet contemplation that can be enjoyed in a wild corner of the city or will watch wildlife in their own garden or terrace. Once suitable habitats are created, species that were once thought to belong in the countryside, far from the city, can colonize. Places like Camley Street Natural Park,[19] the London Wetland Centre[20] and the London Natural History Museum Wildlife Garden,[21] or the biodiverse green roofs of Switzerland have shown what can be achieved. In the water-sensitive cities of the future there will be more and more of these projects and many conventional parks will also be modified to include nature reserves, which also serve an important flood-management function (for example like Sutcliffe Park in the London Borough of Greenwich).[22]

Rediscovering Urban Waterways

City dwellers did turn their backs on water. Pollution during times when streams were used as sewers and a populace where the people with the ability to swim were in a small minority created a common and widespread perception that water is very perilous. Since the 1970s, however, water quality in rivers and coastal waters of developed countries has gradually improved. Most people now learn to swim and our natural childhood fascination with and attraction to water continues. Combining these social factors with the new understanding that bringing water and wetlands back into cities restores valuable ecosystem services means that the restoration of watercourses is now of increasing interest. Projects like the recreation of the Cheonggyecheon River in Seoul, South Korea, which was once covered by a double-decker highway, are inspiring people to look at their own lost rivers and to seek ways of daylighting them. These projects are often long term and

179

may have to be implemented in phases over many decades, but in the future water-sensitive city, more watercourses will be brought to the surface, providing more open space, better flood management and creating more opportunities to be close to nature, with people expecting to be able to go boating or swimming in most towns.

A Greener Looking City

The water-sensitive city of the future, then, will look quite different. In most cities there will be taller buildings but whether they are high rise or lower in stature, many more buildings will have green roofs. Some will be terraces where people relax, other roofs will be vegetable gardens and others left to nature. Buildings will feature large installations of photovoltaic cells. Many facades will be green with planted modules or climbing plants. Vegetation will spill from balconies. Streets will be shadier because of the extra trees and quieter and cleaner as fossil-fuel powered vehicles are replaced with electric- or hydrogen-powered vehicles and cycles. Streets will be cooler and kinder places to be, with the difference between streets and parks less apparent. Parks will bleed into the street. Minor watercourses will be brought back to the surface, rescued from the underground drains that they have been confined to and the larger watercourses will be softer, greener, cleaner and more welcoming.

Living with Climate Change

Cities will need to continue to adapt to climate change. This may mean gradual changes to the vegetation of roofs, walls streets and parks and adjustments to drainage systems to cope with hotter weather and heavier downpours. There will also be the need to adjust to more frequent and deeper inundations of floodplains and rising sea levels. Some parts of cities may need to be rezoned from housing to industry or from industry to open space, depending on the severity of the floods or predicted floods. Other changes will involve changes in building typologies, with buildings raised above flood levels or structures that are able watertight and therefore temporarily able to float for the duration of a flood.[23] Others, like planners in Boston, are proposing the creation of a new podium level in coastal cities, so that roads and transportation routes are elevated above the high water mark at around 10 m in height. Accommodation and workplaces would be at safe, higher levels and the existing ground level would be gradually given over to less vulnerable uses, including sports, recreation and wildlife habitats.[24] In such a city, it would be possible to allow extensive marshes to become established below and between the buildings and main transportation routes.

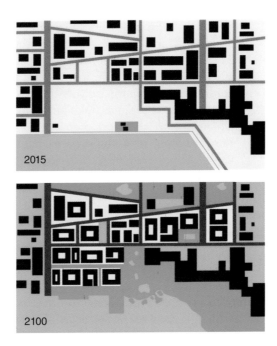

Figure 14.6 Proposal to adapt Boston gradually to a rise in sea level over the coming century. Key roads will be elevated, new raised multistorey blocks built and many existing ground-level roads and buildings will be converted to floodable gardens, parks and wetlands.[24] Illustration by Marianna Magklara based on submission to Boston Living With Water project.

Tough Decisions

Not all cities will be saved. Climate change brought about by the release of greenhouse gases into the atmosphere will lead to a global average temperature increase of at least 2 °C and a rise in sea level of several metres, probably 6 m or more. Storms, floods and drought will overwhelm some cities. Some will not have the capacity to rebuild after catastrophes or to adapt. Whole regions may become too inhospitable to continue to support large conurbations. There will be difficult decisions to be made and relatively safe cities will have to welcome refugees from places where life becomes unbearable.

Younger and Wiser

The inhabitants of future water-sensitive cities will know much more about water than the current generation. They will know how water threads its way through and sustains the urban fabric. They will know that clean freshwater is a scarce resource and they will know about and will take a close interest in where and how their water is sourced. There will always be room for improvement, of course. This will come about through increasing awareness of the need to conserve water and bring soil, water and vegetation into the built environment. Schools will spend more time teaching students about water and students will engage in many more practical and academic projects centred around water. Universities are already increasing their research effort on these subjects (take for example Imperial College's Blue Green Dream)[25] and this and many other initiatives will drive innovation. Knowledge of sustainable water management in future water-sensitive cities will not be confined to specialists alone but will be much more widespread. There will be an effort to ensure that all sectors of society can enjoy a reliable supply of clean water and sanitation as well as easy access to water in the wider environment. People will be thinking about how their homes and workplaces can be secured in the face of extreme weather, including heat waves, drought and deluge. The hope is that these facilities and comforts will not be provided at the expense of the natural environment.

Hope

What this brief review of the issues and opportunities does illustrate, however, is that there is so much waste of water and inefficiency in urban water systems and so many possibilities for improvement that there is reason for hope. There are already many excellent and inspiring examples to follow, including national and city government programmes and grassroots initiatives. Clever techniques and ambitious schemes have been devised to allow cities to adapt to climate change. Helpful direction has been provided. All that is now required is enthusiastic, coordinated and massive action.

Useful Resources

Water Molecule

Stephen Lower of the Simon Fraser University introduces the water molecule, part of his virtual textbook:
http://www.chem1.com/acad/webtext/virtualtextbook.html (accessed 10 November 2015).

Water Cycle

The United States Geological Survey (USGS) has a web-based water science school, which provides this excellent description of the global water cycle:
http://water.usgs.gov/edu/watercycle.html (accessed 10 November 2015).

Ancient Civilizations and Sanitation

Part of a comprehensive history of sanitary sewers, an article by Jon C. Schladweiler, that looks at sanitation in the ancient world:
http://www.sewerhistory.org/time-lines/tracking-down-the-roots-of-our-sanitary-sewers/part-1-the-early-roots-3200-bce-to-300-ce/ (accessed 10 November 2015).

Germ Theory

Harvard University Library's introduction to germ theory, which includes many useful links:
http://ocp.hul.harvard.edu/contagion/germtheory.html (accessed 10 November 2015).

The Water Sensitive City, First Edition. Gary Grant.
© 2016 John Wiley & Sons, Ltd. Published 2016 by John Wiley & Sons, Ltd.

Beginnings of Modern Water Supply and Sanitation

Jon C. Schladweiler's web site again. This time he looks at the origins of modern sanitation:

http://www.sewerhistory.org/time-lines/tracking-down-the-roots-of-our-sanitary-sewers/part-2-the-middle-ages/ (accessed 10 November 2015).

The US EPA has provided this brief history of drinking water treatment:

http://water.epa.gov/aboutow/ogwdw/upload/2001_11_15_consumer_hist.pdf (accessed 11 November 2015).

Water Use

The World Health Organization's advice on the minimum quantity of water required:

http://www.who.int/water_sanitation_health/emergencies/qa/emergencies_qa5/en/ (accessed 10 November 2015).

The USGS on domestic water use in the United States:

http://water.usgs.gov/edu/wudo.html (accessed 10 November 2015).

A poster by the SASI Group (University of Sheffield) and Mark Newman (University of Michigan), on domestic water use across the world:

http://www.worldmapper.org/posters/worldmapper_map324_ver5.pdf (accessed 10 November 2015).

A running tally of water consumption and several useful links. Also interesting tallies of global population, expenditure and energy consumption:

http://www.worldometers.info/water/ (accessed 10 November 2015).

Water Footprinting

Introduction to water footprinting by the Water Footprint Network:

http://waterfootprint.org/en/water-footprint/what-is-water-footprint/ (accessed 10 November 2015).

Links to water footprint statistics:

http://waterfootprint.org/en/resources/water-footprint-statistics/ (accessed 10 November 2015).

Uplands and Glaciers

The 'water tower' function of Tibet:

http://www.futurewater.nl/uk/projects/tibet/ (accessed 10 November 2015).

Mountain snowpack and water supply by the USDA:
  http://www.wcc.nrcs.usda.gov/factpub/sect_2.html (accessed 10 November 2015).
USGS on glaciers:
  http://water.usgs.gov/edu/earthglacier.html (accessed 10 November 2015).
Facts on glaciers from the National Snow and Ice Data Center:
  https://nsidc.org/cryosphere/glaciers/quickfacts.html   (accessed 10 November 2015).

Reservoirs and Dams

National Geographic on reservoirs:
  http://education.nationalgeographic.co.uk/encyclopedia/reservoir/ (accessed 10 November 2015).
The International Commission on Large Dams, explains the role of dams and maintains a list:
  http://www.icold-cigb.net/GB/Dams/role_of_dams.asp/   (accessed 10 November 2015).

Aquifers

The USGS Science School on aquifers:
  http://water.usgs.gov/edu/earthgwaquifer.html   (accessed   10 November 2015).
More on aquifers from Live Science:
  http://www.livescience.com/39625-aquifers.html   (accessed   10 November 2015).
National Geographic again, this time on aquifers:
  http://education.nationalgeographic.co.uk/encyclopedia/aquifer/ (accessed 10 November 2015).
The Water Encyclopedia on the Ogallala Aquifer on the High Plains of the United States:
  http://www.waterencyclopedia.com/Oc-Po/Ogallala-Aquifer.html (accessed 10 November 2015).
The British Geological Survey on aquifers in England and Wales:
  http://www.bgs.ac.uk/products/hydrogeology/aquiferDesignation. html (accessed 10 November 2015).

Nitrates

The World Health Organization on nitrates in drinking water:
  http://www.who.int/water_sanitation_health/dwq/chemicals/ nitratenitrite2ndadd.pdf (accessed 10 November 2015).

The basics of nitrates in drinking water from the US EPA:
  http://water.epa.gov/drink/contaminants/basicinformation/nitrate.cfm (accessed 10 November 2015).
Brian Oram on nitrates and nitrites in drinking water and surface waters:
  http://www.water-research.net/index.php/nitrate (accessed 10 November 2015).
The University of Hertfordshire on the nitrate problem in the United Kingdom:
  http://adlib.everysite.co.uk/adlib/defra/content.aspx?id=000IL3890W.16NTBZ2TSNS231 (accessed 10 November 2015).

Desalination

An overview of desalination from the International Desalination Association:
  http://idadesal.org/desalination-101/desalination-overview/ (accessed 10 November 2015).
The USGS on desalination:
  http://water.usgs.gov/edu/drinkseawater.html (accessed 10 November 2015).
Miriam Balaban's directory of desalination events and resources:
  https://www.desline.com/home.php (accessed 10 November 2015).

Climate Change

The most authoritative source, the Intergovernmental Panel on Climate Change (IPCC):
  http://www.ipcc.ch/ (accessed 10 November 2015).
NASA's climate-change site:
  http://climate.nasa.gov/ (accessed 10 November 2015).
The Mayor of London on the city's urban heat island:
  http://legacy.london.gov.uk/mayor/environment/climate-change/docs/UHI_summary_report.pdf (accessed 10 November 2015).

Urban Heat Islands

The US EPA on urban heat islands:
  http://www2.epa.gov/heat-islands (accessed 10 November 2015).
Camilio Perez Arrau and Marco A. Pena have provided this useful links on urban heat islands:
  http://www.urbanheatislands.com/external-links (accessed 10 November 2015).

Microclimate

The UK Met Office fact sheet on microclimates:
http://www.metoffice.gov.uk/media/pdf/n/9/Fact_sheet_No._14.pdf (accessed 10 November 2015).
The British Geographer on urban climates:
http://thebritishgeographer.weebly.com/urban-climates.html (accessed 10 November 2015).

Ecosystem Approach

The Convention on Biological Diversity on the ecosystem approach
https://www.cbd.int/ecosystem/principles.shtml (accessed 10 November 2015).
The Millennium Ecosystem Assessment was called for by the UN Secretary General in 2000:
http://www.millenniumassessment.org/en/index.html (accessed 10 November 2015).
The Ecosystem Knowledge Network is the United Kingdom's leading source on the ecosystem approach:
http://ecosystemsknowledge.net/ (accessed 10 November 2015).

Rivers

The European Centre for River Restoration champions urban river restoration across Europe:
http://www.ecrr.org/RiverRestoration/UrbanRiverRestoration/tabid/3177/Default.aspx (accessed 10 November 2015).
The River Restoration Centre is the United Kingdom's leading source of information on river restoration:
http://www.therrc.co.uk/ (accessed 10 November 2015).
Adrian Beneppe on urban stream restoration and daylighting:
http://www.thenatureofcities.com/2013/10/30/up-the-creek-with-a-paddle-urban-stream-restoration-and-daylighting/ (accessed 10 November 2015).
National Geographic on long-buried urban streams seeing the light again:
http://news.nationalgeographic.com/news/2014/11/141125-dc-daylighting-broad-branch-stream-restoration-science/ (accessed 10 November 2015).
American Rivers on restoring urban rivers:
http://www.americanrivers.org/initiatives/urban-rivers/ (accessed 10 November 2015).

American Trails on urban river parkways:

http://www.americantrails.org/resources/health/Urban-River-Parkways-health.html (accessed 10 November 2015).

Near-Natural Drainage

Low impact development (LID) explained:

http://www.lowimpactdevelopment.org/ (accessed 10 November 2015).

LID tools:

http://www.lid-stormwater.net/ (accessed 10 November 2015).

Sustainable drainage information provided by the Construction Industry Research Information Association:

http://www.susdrain.org/ (accessed 10 November 2015).

Rain Gardens

The City of Portland, Oregon has provided this advice on creating rain gardens:

http://www.portlandoregon.gov/bes/article/337963 (accessed 10 November 2015).

North Carolina State University:

https://www.bae.ncsu.edu/topic/raingarden/Building.htm (accessed 10 November 2015).

The UK rain garden guide, written by the author:

http://raingardens.info/ (accessed 10 November 2015).

Water-Sensitive Cities

This Australian consortium of government departments and agencies, water companies and universities has provided this comprehensive source of information on the water-sensitive city:

http://watersensitivecities.org.au/ (accessed 10 November 2015).

Water Conservation

More than 100 ways of conserving water, provided by a coalition of cities in Arizona:

http://wateruseitwisely.com/100-ways-to-conserve/ (accessed 10 November 2015).

Ways of conserving water at home:

http://eartheasy.com/live_water_saving.htm (accessed 10 November 2015).

The *National Geographic*'s water conservation tips:
   http://environment.nationalgeographic.com/environment/freshwater/water-conservation-tips/ (accessed 10 November 2015).
Waterwise on saving water:
   http://www.waterwise.org.uk/pages/how-to-save-water.html (accessed 10 November 2015).

Rainwater Harvesting

Waterwise on rainwater harvesting:
   http://www.waterwise.org.uk/pages/rainwater-harvesting.html (accessed 10 November 2015).
The American Rainwater Catchment Systems Association promotes sustainable rainwater harvesting practices:
   http://www.arcsa.org/ (accessed 10 November 2015).
The Rainwater Harvesting Association of Australia promotes rainwater harvesting and is a useful source of information:
   http://www.rainwaterharvesting.org.au/newsmedia/publications (accessed 11 November 2015).
Rainwater harvesting in India:
   http://www.waterharvesters.com/rain-water-harvesting (accessed 10 November 2015).

Greywater Recycling

Greywater Action is a group of educators looking to reduce water use. Here is their advice on greywater recycling:
   http://greywateraction.org/contentabout-greywater-reuse/ (accessed 10 November 2015).
EPA on water recycling – makes reference to greywater:
   http://www3.epa.gov/region09/water/recycling/ (accessed 10 November 2015).
The *Guardian* on greywater systems:
   http://www.theguardian.com/lifeandstyle/2014/jul/21/greywater-systems-can-they-really-reduce-your-bills (accessed 10 November 2015).

Green Roofs and Living Walls

The US site includes the international green roofs and living walls project database:
   http://www.greenroofs.com/ (accessed 10 November 2015).
The UK site for advice on green roofs:
   http://www.livingroofs.org/ (accessed 10 November 2015).

Staffordshire University's green wall centre:
http://www.staffs.ac.uk/research/greenwall/#what are green walls?
(accessed 10 November 2015).

Street Trees and Water

EPA's useful guide to managing stormwater with street trees:
http://water.epa.gov/polwaste/green/upload/stormwater2
streettrees.pdf (accessed 10 November 2015).
Presentation on work to improve tree pits in Stockholm:
http://www.ikt.de/website/forum2013/oestberg.pdf (accessed 10
November 2015).

Urban Agriculture

The Food and Agriculture Organization's guide to urban agriculture:
http://www.fao.org/urban-agriculture/en/

Urban Biodiversity

UNEP's introduction to urban biodiversity:
http://www.unep.org/urban_environment/issues/biodiversity.asp
(accessed 10 November 2015).
UNESCO on 'urban systems':
http://www.unesco.org/new/en/natural-sciences/environment/
ecological-sciences/specific-ecosystems/urban-systems/ (accessed 10
November 2015).
Local Governments for Sustainability, has the URBES project, which
looks at urban biodiversity and ecosystem services:
http://cbc.iclei.org/about-urbes (accessed 10 November 2015).
Discovering urban biodiversity in New York City:
http://www.thenatureofcities.com/2012/08/14/discovering-urban-
biodiversity/ (accessed 10 November 2015).
The Greater London Authority on London's biodiversity, including
many interesting links:
https://www.london.gov.uk/priorities/environment/greening-
london/biodiversity (accessed 10 November 2015).

Notes

Chapter 1

1. http://www1.lsbu.ac.uk/water/life.html (accessed 2 November 2015).
2. http://ga.water.usgs.gov/edu/earthhowmuch.html (accessed 28 October 2015).
3. http://www.drinking-water.org/html/en/glossary.html#gloss59 (accessed 28 October 2015).
4. http://www.ucmp.berkeley.edu/glossary/gloss5/biome/ (accessed 28 October 2015).
5. http://www.theguardian.com/environment/2012/mar/22/water-wars-countries-davey-warns (accessed 28 October 2015)
6. http://www.elnino.noaa.gov/ (accessed 28 October 2015).
7. http://www.worldclimate.com/ (accessed 28 October 2015).
8. http://www.elnino.noaa.gov/ (accessed 28 October 2015).
9. http://www.laht.com/article.asp?ArticleId=347191&CategoryId=14089 (accessed 28 October 2015).
10. http://www.ncdc.noaa.gov/paleo/pubs/ray2001/ray2001.html (accessed 28 October 2015).
11. http://co2now.org/Current-CO2/CO2-Now/ (accessed 28 October 2015).
12. Molden, D. (Ed) (2007) *Water for Food, Water for Life: A Comprehensive Assessment of Water Management in Agriculture*, Earthscan/IWMI, London, p.2.
13. http://earthobservatory.nasa.gov/Features/WorldOfChange/aral_sea.php (accessed 28 October 2015).
14. http://www.fao.org/docrep/x5871e/x5871e00.htm (accessed 28 October 2015).
15. http://centaur.reading.ac.uk/3339/ (accessed 28 October 2015).
16. http://www.waterhistory.org/histories/rome/ (accessed 28 October 2015).
17. http://books.google.co.uk/books?id=eLA4TnmQqzAC&lpg=PR1&pg=PA1#v=onepage&q&f=false (accessed 28 October 2015).
18. http://www.worldenergyoutlook.org/resources/water-energynexus/?utm_content=buffere5911&utm_medium=social&utm_source=twitter.com&utm_campaign=buffer
19. http://www.nexuswaterenergy.com/ (accessed 28 October 2015).
20. http://www.camdennewjournal.co.uk/archive/r111203_4.htm (accessed 28 October 2015).
21. http://www.choleraandthethames.co.uk/cholera-in-london/the-great-stink/ (accessed 28 October 2015).

22. http://www.choleraandthethames.co.uk/cholera-in-london/the-big-thames-clean-up/sir-joseph-bagalgette/ (accessed 28 October 2015).
23. http://www.iied.org/water-sanitation-0 (accessed 28 October 2015).
24. http://earthobservatory.nasa.gov/IOTD/view.php?id=44677 (accessed 28 October 2015).
25. http://yosemite.epa.gov/water/owrccatalog.nsf/065ca07e299b4646852 56ce50075c11a/385eafdb27690f9f85256efc006824a9!OpenDocument (accessed 2 November 2015).
26. http://water.epa.gov/polwaste/green/ (accessed 28 October 2015).
27. http://www.susdrain.org/ (accessed 28 October 2015).
28. http://www.data360.org/dsg.aspx?Data_Set_Group_Id=757 (accessed 28 October 2015).
29. http://www.waterwise.org.uk/data/resources/25/Water_factsheet_2012. pdf (accessed 28 October 2015).
30. Stedman, L. (2006) Motivations for metering. *Water 21* (April) pp. 26–28.
31. http://www.eea.europa.eu/publications/Sustainableuseofwater (accessed 28 October 2015).
32. http://www.rainwaterharvesting.org/international/china.htm (accessed 28 October 2015).
33. http://greywateraction.org/contentabout-greywater-reuse/ (accessed 2 November 2015).
34. http://www.pub.gov.sg/water/newater/Pages/default.aspx (accessed 28 October 2015).
35. http://www.millenniumassessment.org/en/index.html (accessed 2 November 2015).
36. http://www.cbd.int/gbo3/ (accessed 28 October 2015).
37. http://www.fao.org/fileadmin/templates/FCIT/PDF/Fact_sheet_on_ aquaponics_Final.pdf (accessed 28 October 2015).
38. http://www.guerrillagardening.org/ (accessed 28 October 2015).
39. http://www.ibm.com/smarterplanet/us/en/smarter_cities/overview/ (accessed 28 October 2015).
40. http://www.melbournewater.com.au/wsud (accessed 28 October 2015).

Chapter 2

1. http://waterbornepathogens.susana.org/ (accessed 2 November 2015).
2. Jansen, M. (1989) Water supply and sewage disposal at Mohenjo-daro. *World Archaeology* **21**(2), http://www.personal.leeds.ac.uk/~cen6ddm/ History/Mojenjodaro2.pdf (accessed 28 October 2015).
3. http://oi.uchicago.edu/pdf/oip24.pdf (accessed 28 October 2015).
4. Dalley, S. (2013) *The Mystery of the Hanging Garden of Babylon: An Elusive World Wonder Traced*, Oxford University Press, New York.
5. http://www.academia.edu/543256/Sherds_clay_and_clean_Water_-_ Ancient_Egyptian_Well-technology (accessed 28 October 2015).
6. Castleden, R. (1990) *The Knossos Labyrinth*, Routledge, London.
7. http://www.edwardgoldsmith.org/1031/the-qanats-of-iran/ (accessed 28 October 2015).
8. http://www.pompeiiinpictures.com/pompeiiinpictures/Fountains/ fountain%2061500.htm (accessed 2 November 2015).
9. http://www.shca.ed.ac.uk/projects/longwalls/WaterSupply.htm (accessed 28 October 2015).

10. Çeçen, K. (1996) *The Longest Roman Water Supply Line*, TSKB, Istanbul.
11. http://www.waterhistory.org/histories/aguadas/ (accessed 28 October 2015).
12. http://www.sciencedaily.com/releases/2012/02/120223142455.htm (accessed 28 October 2015).
13. http://www.asce.org/project/tipon/ (accessed 2 November 2015).
14. http://whc.unesco.org/en/list/1001 (accessed 28 October 2015).
15. http://whc.unesco.org/en/list/811 (accessed 28 October 2015).
16. Johnson, S. (2006) *The Ghost Map: The Story of London's Most Terrifying Epidemic – and How it Changed Science, Cities and the Modern World*, Riverhead Books, New York.
17. http://www.sciencemuseum.org.uk/broughttolife/people/maryseacole.aspx (accessed 28 October 2015).
18. http://www.bbc.co.uk/timelines/z92hsbk (accessed 28 October 2015).
19. http://www.sciencemuseum.org.uk/broughttolife/techniques/germtheory.aspx (accessed 28 October 2015).
20. Thames Water (n.d.) *London's Victorian Sewer System*, https://www.thameswater.co.uk/about-us/10092.htm (accessed 28 October 2015).
21. http://www.hamburgwatercycle.de/index.php/conventional-wastewater-treatment.html (accessed 28 October 2015).
22. Burian, S. J., Nix, S. J., Pitt, R. E. and Durrans, S. R. (2000) Urban wastewater management in the United States: past, present, and future. *Journal of Urban Technology* **7**(3), 33–62, http://www.sewerhistory.org/articles/whregion/urban_wwm_mgmt/urban_wwm_mgmt.pdf (accessed 28 October 2015).
23. Frankland, E. (1880) *Water Analysis for Sanitary Purposes*, J. van Voorst, London.
24. Seeger, H. (1999) The history of German wastewater treatment. *European Water Management* **2**(5), 51–56.
25. http://www.asce.org/project/lawrence-experimental-station/ (accessed 2 November 2015).
26. Royal Commission on Sewage Disposal (1915) *Final Report on Treating and Disposal of Sewage*, HMSO, London, http://ia700404.us.archive.org/35/items/cu31924003641929/cu31924003641929.pdf (accessed 28 October 2015).
27. http://www.iwa100as.org/history.php (accessed 2 November 2015).
28. http://history.powys.org.uk/history/rhayader/elanmenu.html (accessed 28 October 2015).
29. Kahrl, W. L. (1982) *Water and Power: The Conflict over Los Angeles' Water Supply in the Owens Valley*, University of California Press, Berkeley, CA.
30. http://www.ovcweb.org/issues/desertification.html (accessed 28 October 2015).
31. http://www.wunderground.com/news/california-exceptional-drought-first-time-history-20140130 (accessed 28 October 2015).

Chapter 3

1. http://www.sphereproject.org/ (accessed 28 October 2015).
2. http://www.who.int/water_sanitation_health/publications/2011/WHO_TN_09_How_much_water_is_needed.pdf?ua=1 (accessed 28 October 2015).

3. http://water.usgs.gov/edu/qa-home-percapita.html (accessed 28 October 2015).

4. http://www.epa.gov/WaterSense/pubs/indoor.html (accessed 28 October 2015).

5. http://www.waterwise.org.uk/data/resources/25/Water_factsheet_2012.pdf (accessed 29 October 2015).

6. https://www.anglianwater.co.uk/_assets/media/Fact_File_5_-_Using_ water_at_home.pdf (accessed 29 October 2015).

7. German Federal Ministry for the Environment Nature Conservation, Building and Nuclear Safety (n.d.) *Drinking Water*, http://www.bmub.bund.de/en/topics/water-waste-soil/water-management/drinking-water/ (accessed 29 October 2015).

8. Wirtschafts und Verlagsgesellschaft Gas und Wasser mbH (2008) *Profile of the German Water Industry 2008*, http://www.bdew.de/bdew.nsf/id/DE_Profile_of_the_German_Water_Industry/$file/Profile_German_Water_Industry_2008.pdf (accessed 29 October 2015).

9. Hoekstra, A. Y., Chapagain, A. K., Aldaya, M. M and Mekonnen, M. M. (2011) *The Water Footprint Assessment Manual: Setting the Global Standard*, Earthscan, London, http://waterfootprint.org/media/downloads/TheWaterFootprintAssessmentManual_2.pdf (accessed 2 November 2015).

10. http://hdr.undp.org/en/media/HDR06-complete.pdf (accessed 29 October 2015).

11. http://www.internationalrivers.org/campaigns/south-north-water-transfer-project (accessed 29 October 2015).

12. http://www.theguardian.com/environment/chinas-choice/2014/apr/18/china-one-fifth-farmland-soil-pollution (accessed 29 October 2015).

13. Sonnenberg, A., Chapagain, A., Geiger, M and August, D. (2009) *Der Wasser-Fussabdruck Deutchlands: Woher stamillimetrest das Wasser das in unseren Lebensmitteln steckt?* WWF Deutschland, Frankfurt, http://www.waterfootprint.org/Reports/Sonnenberg-et-al-2009-Wasser-Fussabdruck-Deutschlands.pdf (accessed 29 October 2015).

14. Kampman, D. A., Hoekstra, A. Y. and Krol, M. S. (2008) *The Water Footprint of India*. Value of Water Research Report Series No.32, UNESCO-IHE, Delft.

15. http://www.waterfootprint.org/Reports/Report37-WaterFootprint-Indonesia.pdf (accessed 29 October 2015).

16. http://www.crisp.nus.edu.sg/~acrs2001/pdf/126boehm.pdf (accessed 29 October 2015).

17. http://www.waterfootprint.org/Reports/Aldaya-el-al-2010-Water FootprintSpain.pdf (accessed 29 October 2015).

18. http://www.waterfootprint.org/?page=files/WaterStat-Product WaterFootprints (accessed 29 October 2015).

19. http://pubs.usgs.gov/circ/1344/pdf/c1344.pdf (accessed 29 October 2015).

20. Mekonnen, M. M. and Hoekstra, A. Y. (2011) The green, blue and grey water footprint of crops and derived crop products. *Hydrology and Earth System Sciences* **15**, 1577–1600, http://www.waterfootprint.org/Reports/Mekonnen-Hoekstra-2011-WaterFootprintCrops.pdf (accessed 29 October 2015).

21. Gutentag, E. D., Heimes, F. J., Krothe, N. C., Luckey, R. R. and Weeks, J. B. (1984) *Geohydrology of the High Plains Aquifer in parts of Colorado, Kansas, Nebraska, New Mexico, Oklahoma, South Dakota, Texas, and*

Wyoming. USGS professional paper: 1400-B, http://pubs.er.usgs.gov/publication/pp1400B (accessed 29 October 2015).

22. *The Ogallala Aquifer Initiative*, http://www.nrcs.usda.gov/wps/portal/nrcs/detailfull/national/programs/initiatives/?cid=stelprdb2048809 (accessed 29 October 2015).

23. http://energy.sandia.gov/climate-earth-systems/water-security-program/energywater-history/ (accessed 2 November 2015).

24. http://www.ucsusa.org/clean_energy/our-energy-choices/energy-and-water-use/water-energy-electricity-cooling-power-plant.html#.VGi2qZOsUac (accessed 29 October 2015).

25. ftp://ftp.cordis.europa.eu/pub/estep/docs/water-steel-report_en.pdf (accessed 29 October 2015).

26. https://www.worldsteel.org/dms/internetDocumentList/bookshop/2015/Water_position_paper_2015_vfinal/document/Position%20paper%20Water%20management%20in%20the%20steel%20industry.pdf (accessed 2 November 2015).

27. http://www.lcm2011.org/papers.html?file=tl_files/pdf/poster/day2/Horie-Comparison_of_Water_Footprint_for_Industrial_Products-566_b.pdf (accessed 2 November 2015).

28. Gleick, P. H. (1994) 'Water and energy' in annual reviews. *Annual Review of Energy and the Environment* **19**, 267–299.

29. http://energy.gov/sites/prod/files/2013/03/f0/ShaleGasPrimer_Online_4-2009.pdf (accessed 2 November 2015).

30. http://www.sourcewatch.org/index.php/Fracking_and_water_consumption (accessed 29 October 2015).

Chapter 4

1. http://www.unep.org/geo/geo_ice/PDF/full_report_LowRes.pdf (accessed 29 October 2015).

2. http://www.ruralpovertyportal.org/country/voice/tags/kenya/mount_kenya (accessed 29 October 2015).

3. http://www.fao.org/docrep/018/i3364e/i3364e02.pdf (accessed 29 October 2015).

4. Burger, H. (1954) Einfluss des Waldes auf den Stand der Gewässer – der Wasserhaushalt im Sperbel- und Rappengraben von 1944/45 bis 1953/54. *Mitteilungen EAFV* [Birmensdorf] **31**(2), 493–555.

5. http://www.nyc.gov/html/dep/html/watershed_protection/east_branch.shtml (accessed 29 October 2015).

6. http://www.engineering-timelines.com/scripts/engineeringItem.asp?id=1292 (accessed 29 October 2015).

7. http://www.engineering-timelines.com/scripts/engineeringItem.asp?id=1315 (accessed 29 October 2015).

8. http://www.michigan.gov/dnr/0,4570,7-153-10364_52259_19092-46291--,00.html (accessed 29 October 2015).

9. https://www.researchgate.net/publication/228713212_Coastal_erosion_and_habitat_loss_along_the_Godavari_delta_front-_a_fallout_of_dam_construction_%28%29 (accessed 29 October 2015).

10. http://www.academia.edu/6520578/Stratification_in_Reservoirs_and_Lakes (accessed 29 October 2015).

11. http://www.internationalrivers.org/human-impacts-of-dams (accessed 29 October 2015).
12. https://www.gov.uk/government/uploads/system/uploads/attachment_data/file/297309/LIT_4892_20f775.pdf (accessed 29 October 2015).
13. https://www.gov.uk/nitrate-vulnerable-zones (accessed 29 October 2015).
14. http://www.euwfd.com/html/groundwater.html (accessed 29 October 2015).
15. http://water.usgs.gov/edu/gwdepletion.html (accessed 29 October 2015).
16. http://blogs.ei.columbia.edu/2010/09/16/cyprus-a-case-study-in-water-challenges/ (accessed 29 October 2015).
17. http://www.globalwaterintel.com/desalination-industry-enjoys-growth-spurt-scarcity-starts-bite/ (accessed 29 October 2015).
18. http://www.sidem-desalination.com/en/Process/ (accessed 29 October 2015).
19. http://puretecwater.com/what-is-reverse-osmosis.html (accessed 29 October 2015).
20. http://www.pca-gmbh.com/appli/desalen.htm (accessed 29 October 2015).
21. https://www.ofwat.gov.uk/consumerissues/rightsresponsibilities/waterpressure (accessed 29 October 2015).
22. http://www.who.int/water_sanitation_health/hygiene/plumbing10.pdf (accessed 29 October 2015).
23. http://water.epa.gov/drink/contaminants/ (accessed 29 October 2015).

Chapter 5

1. http://www.meteo.psu.edu/holocene/public_html/shared/articles/littleiceage.pdf (accessed 30 October 2015).
2. http://blogs.scientificamerican.com/plugged-in/why-we-know-about-the-greenhouse-gas-effect/ (accessed 30 October 2015).
3. http://www.aip.org/history/climate/co2.htm#SC (accessed 30 October 2015).
4. http://scrippsco2.ucsd.edu/ (accessed 30 October 2015).
5. http://www.ipcc.ch/organization/organization.shtml (accessed 30 October 2015).
6. http://www.lse.ac.uk/GranthamInstitute/ (accessed 30 October 2015).
7. http://www.ipcc.ch/index.htm (accessed 30 October 2015).
8. http://climate.nasa.gov/evidence/ (accessed 30 October 2015).
9. http://rsta.royalsocietypublishing.org/content/369/1934/6 (accessed 30 October 2015).
10. Church, J. A. and White, N. J. (2006) A twentieth century acceleration in global sea level rise. *Geophysical Research Letters* **33**, L01602, doi:10.1029/2005GL024826.
11. http://www.copenhagendiagnosis.com/read/default.html (accessed 30 October 2015).
12. http://www.smh.com.au/environment/un-climate-conference/paris-2015-two-degrees-warming-a-prescription-for-disaster-says-top-climate-scientist-james-hansen-20150504-ggu33w (accessed 30 October 2015).
13. http://www.nola.com/katrina/index.ssf/2013/08/hurricane_katrina_eight_years.html (accessed 30 October 2015).

14. http://www.wired.com/2015/02/rising-sea-levels-already-making-miamis-floods-worse/ (accessed 30 October 2015).

15. http://www.oco.noaa.gov/resources/Documents/Levitus_GRL_Heat2008.pdf (accessed 2 November 2015).

16. http://www.ncdc.noaa.gov/extremes/cei/index.html (accessed 30 October 2015).

17. http://climate.nasa.gov/vital-signs/land-ice/ (accessed 30 October 2015).

18. http://climate.nasa.gov/interactives/global_ice_viewer (accessed 30 October 2015).

19. http://www.the-cryosphere.net/9/269/2015/tc-9-269-2015.html (accessed 30 October 2015).

20. http://climate.nasa.gov/vital-signs/arctic-sea-ice/ (accessed 30 October 2015).

21. http://www.independent.co.uk/environment/climate-change/exclusive-the-methane-time-bomb-938932.html (accessed 30 October 2015).

22. https://robertscribbler.wordpress.com/2015/06/01/arctic-methane-alert-ramp-up-at-numerous-reporting-stations-shows-signature-of-an-amplifying-feedback/ (accessed 30 October 2015).

23. http://www.pmel.noaa.gov/co2/story/What+is+Ocean+Acidification%3F (accessed 2 November 2015).

24. http://www.nrdc.org/water/files/ca-snowpack-and-drought-FS.pdf (accessed 30 October 2015).

25. http://ukclimateprojections.metoffice.gov.uk/21678 (accessed 30 October 2015).

26. http://ukclimateprojections.metoffice.gov.uk/22530 (accessed 30 October 2015).

27. http://www.nature.com/nclimate/journal/v4/n7/full/nclimate2258.html (accessed 30 October 2015).

28. http://www.rferl.org/content/penguin-from-flooded-tbilisi-zoo-reaches-azerbaijani-border/27077411.html (accessed 30 October 2015).

29. http://floodlist.com/asia/tbilisi-floods-death-toll-rises-eu-aid-georgia (accessed 30 October 2015).

30. http://www.metoffice.gov.uk/learning/learn-about-the-weather/weather-phenomena/heatwave (accessed 30 October 2015).

31. http://www.ncbi.nlm.nih.gov/pubmed/18235058 (accessed 30 October 2015).

32. http://www.metoffice.gov.uk/learning/learn-about-the-weather/weather-phenomena/case-studies/russianheatwave (accessed 30 October 2015).

33. https://www.climatecommunication.org/new/features/heat-waves-and-climate-change/heat-waves-the-details/#refmark-2 (accessed 30 October 2015).

34. http://www.drought.gov/drought/ (accessed 30 October 2015).

35. http://www.drought.gov/gdm/current-conditions (accessed 30 October 2015).

36. http://www.metoffice.gov.uk/nhp/media.jsp?mediaid=15102&filetype=pdf (accessed 2 November 2015).

37. http://climatemigration.org.uk/moving-stories-the-sahel/ (accessed 30 October 2015).

Chapter 6

1. http://www.climate-charts.com/index.html (accessed 30 October 2015).
2. http://www.metoffice.gov.uk/media/pdf/n/9/Fact_sheet_No._14.pdf (accessed 30 October 2015).
3. http://www2.epa.gov/heat-islands (accessed 2 November 2015).
4. http://news.nationalgeographic.com/news/energy/2013/10/131022-harbin-ice-city-smog-crisis-china-coal/ (accessed 30 October 2015).
5. http://www.theguardian.com/environment/2014/mar/25/air-pollution-single-biggest-environmental-health-risk-who (accessed 30 October 2015).
6. http://www.sciencedirect.com/science/article/pii/S1352231015000758 (accessed 30 October 2015).
7. http://www.gebaeudekuehlung.de/URS2009Marseille.pdf (accessed 30 October 2015).
8. http://www.whiteroofproject.org/ (accessed 30 October 2015).
9. http://www.academia.edu/6649534/Living_Walls_more_than_scenic_beauties (accessed 30 October 2015).
10. McPherson, E. G. (1993) Evaluating the cost effectiveness of shade trees for demand-side management. *Electricity Journal* **6**, 57–65.
11. http://www.forestry.gov.uk/pdf/FCRN012.pdf/$FILE/FCRN012.pdf (accessed 30 October 2015).
12. https://www.researchgate.net/publication/227882360_Microclimate_modelling_of_street_tree_species_effects_within_the_varied_urban_morphology_in_the_Mediterranean_city_of_Tel_Aviv_Israel (accessed 30 October 2015).
13. Saito, I. (1990–1) Study of the effect of green areas on the thermal environment in an urban area. *Energy and Buildings* **15**, 493–498
14. Jauregui, E. (1990–1) Influence of a large urban park on temperature and convective precipitation in a tropical city. *Energy and Buildings* **15**, 457–463.
15. http://www.sciencedirect.com/science/article/pii/S0378778898000322 (accessed 30 October 2015).
16. Givoni, B. (1998) Impact of green areas on site and urban climates. In *Climate Considerations in Building and Urban Design* (ed. B Givoni). John Wiley & Sons, Inc., New York, pp. 303–330.
17. Meier, A. K. (1990–1) Strategic landscaping and air conditioning savings: A literature review. *Energy and Buildings* **15**, 479–486.
18. Coutts, A. M., Tapper, N. J., Beringer, J., Loughnan, M. and Demuzere, M. (2013) Watering our cities: the capacity for water sensitive urban design to support urban cooling and improve human thermal comfort in the Australian context. *Progress in Physical Geography* **37**(1), 2–28.
19. Murakawa, S., Sekine, T. and Narita, K. (1990) Study of the effects of a river on the thermal environment in an urban area. *Energy Build* **15**, 993–1001.
20. Department of Health (2008) *Heatwave Plan for England. Protecting Health and Reducing Harm from Extreme Heat and Heatwaves.* NHS/Department of Health, London.
21. http://www.firstpost.com/india/mega-heat-wave-kills-over-1400-with-majority-in-andhra-pradesh-telangana-the-bad-news-is-it-will-get-worse-2264580.html (accessed 30 October 2015).
22. http://www.cityofchicago.org/city/en/depts/dgs/supp_info/city_hall_green_roof.html (accessed 2 November 2015).
23. Liu, K. Y. and Baskaran, A. (2005) National Research Council of Canada, http://www.nrc-cnrc.gc.ca/ctu-sc/ctu_sc_n65 (accessed 2 November 2015).

Chapter 7

1. http://www.igbp.net/globalchange/greatacceleration.4.1b8ae20512db6 92f2a680001630.html (accessed 30 October 2015).
2. http://hmpdacc.org/ (accessed 30 October 2015).
3. http://news.stanford.edu/news/2014/july/sixth-mass-extinction-072414. html (accessed 30 October 2015).
4. http://www.cbd.int/ (accessed 30 October 2015).
5. http://www.cbd.int/cop12/ (accessed 30 October 2015).
6. http://www.cbd.int/ecosystem/principles.shtml (accessed 30 October 2015).
7. http://jncc.defra.gov.uk/default.aspx?page=6380 (accessed 30 October 2015).
8. http://www.teebweb.org/resources/ecosystem-services/ (accessed 30 October 2015).
9. http://www.teebweb.org/ (accessed 30 October 2015).
10. http://forestry.usu.edu/htm/city-and-town/urbancommunity-forestry/ what-is-a-tree-worth (accessed 2 November 2015).
11. http://www.rightlivelihood.org/suzuki_speech.html (accessed 30 October 2015).

Chapter 8

1. http://gis.nyc.gov/doitt/nycitymap/ (accessed 30 October 2015).
2. http://www.oxfordbibliographies.com/view/document/obo-9780199756384/obo-9780199756384-0074.xml (accessed 30 October 2015).
3. http://www.riverlifepgh.org/riverfront-projects/ (accessed 30 October 2015).
4. http://www.clydewaterfront.com/our-journey/image-galleries/doors-open-day-2012 (accessed 30 October 2015).
5. http://www.gateshead.gov.uk/Leisure%20and%20Culture/attractions/ bridge/Backgroundnew.aspx (accessed 30 October 2015).
6. http://www.hafencity.com/en/home.html (accessed 30 October 2015).
7. http://www.environmental-expert.com/Files%5C19643%5Carticles% 5C17311%5CPointFraser.pdf (accessed 30 October 2015).
8. http://www.atkinsglobal.com/~/media/Files/A/Atkins-Global/ Attachments/sectors/water/library-docs/technical-papers/delivering-wetland-biodiversity-in-the-London2012-olympic-Park.pdf (accessed 30 October 2015).
9. http://queenelizabetholympicpark.co.uk/the-park/plan-your-visit/interactive-park-map (accessed 2 November 2015).
10. http://minneapolisriverfrontdesigncompetition.com/teams/winning-team-tlskva (accessed 30 October 2015).
11. http://evidence.environment-agency.gov.uk/FCERM/en/SC060065/ MeasuresList/M7/M7T2.aspx?pagenum=2 (accessed 30 October 2015).
12. http://www.ultimatehistoryproject.com/coney-island.html (accessed 30 October 2015).
13. http://www.piers.org.uk/ (accessed 30 October 2015).

14. http://www.centreforsocialjustice.org.uk/publications/turning-the-tide-social-justice-in-five-seaside-towns (accessed 30 October 2015).

15. http://coastalcare.org/educate/shoreline-engineering/ (accessed 30 October 2015).

Chapter 9

1. http://water.epa.gov/polwaste/nps/archives/upload/42issue.pdf (accessed 30 October 2015).

2. http://www.aacounty.org/DPW/Highways/Resources/Raingarden/RG_Bioretention_PG%20CO.pdf (accessed 2 November 2015).

3. http://www.lowimpactdevelopment.org/about.htm (accessed 30 October 2015).

4. https://www.sepa.org.uk/regulations/water/diffuse-pollution/diffuse-pollution-in-the-urban-environment/ (accessed 2 November 2015).

5. http://uk.practicallaw.com/6-521-3041 (accessed 30 October 2015).

6. http://www.ciria.org/Resources/Free_publications/the_suds_manual.aspx (accessed 30 October 2015).

7. http://www.naturnahe-regenwasserbewirtschaftung.info/index.php?page=home (accessed 30 October 2015).

8. Forschungsgesellschaft Landschaftsentwicklung und Landschaftsbau e. V (2002) *Guidelines for the Planning, Execution and Upkeep of Green Roof Sites*, FLL, Bonn.

9. http://www.nfrc.co.uk/media-centre/nfrc-news-detail/2014/10/29/gro-code-2014-unveiled (accessed 30 October 2015).

10. Stovin, V., Vesuviano, G. and Kasmin, H. (2012) The hydrological performance of a green roof test bed under UK climatic conditions. *Journal of Hydrology* **414–415**, 148–161.

11. Martin, B. K. (2008) *The Dynamic Response of a Green Roof*, University of Guelph, Guelph.

12. Forschungsgesellschaft Landschaftsentwicklung und Landschaftsbau e. V (2002) *Guidelines for the Planning, Execution and Upkeep of Green Roof Sites*, FLL, Bonn.

13. https://www.portlandoregon.gov/bes/34598 (accessed 30 October 2015).

14. https://www.portlandoregon.gov/bes/article/63096 (accessed 30 October 2015).

15. http://www.southportland.org/files/4713/6155/3563/RainGarden.pdf (accessed 30 October 2015).

16. http://www.mrsc.org/subjects/pubworks/sidew.aspx#authority (accessed 30 October 2015).

17. http://ttfwatershed.org/vernon-park/ (accessed 30 October 2015).

18. http://water.epa.gov/polwaste/green/upload/pavements.pdf (accessed 30 October 2015).

19. http://www.baunat.boku.ac.at/iblb/eigene-entwicklungen/boku-schotterrasen/ (accessed 30 October 2015).

20. http://www.waverley.nsw.gov.au/__data/assets/pdf_file/0004/60790/Tree_Root_Damage_to_Pipes_Study_-_Mark_Hartley,_Arborist_Network2012.pdf (accessed 30 October 2015).

21. http://www.asla.org/meetings/am2007/Handouts/MONC8Alt%20to%20Struc%20soil.doc.pdf (accessed 2 November 2015).

22. https://wiki.epa.gov/watershed2/index.php/Urban_tree_planting_in_Stockholm (accessed 30 October 2015).

23. http://www.floatingislandinternational.com/products/ftw-a-deeper-understanding/ (accessed 30 October 2015).

24. http://www.floatingislandinternational.com/wp-content/plugins/fii/casestudies/23.pdf (accessed 30 October 2015).

25. Bulc, G. T., Krivograd-Klemenčič, A. and Razinger, J. (2011) Vegetated ditches for treatment of surface water with highly fluctuating water regime. *Water Science and Technology* **63**(10), 2353–2359.

Chapter 10

1. http://news.nationalgeographic.com/news/2011/03/pictures/110322-world-water-day-top-cities/ (accessed 30 October 2015).

2. http://www.parliament.uk/briefing-papers/SN01509.pdf (accessed 2 November 2015).

3. http://www.energysavingwarehouse.co.uk/learning-portal/cistern-displacement-devices/ (accessed 30 October 2015).

4. http://www.duravit.co.uk/website/homepage/duravit_green/duravit_green/sustainable_products.com-en.html (accessed 30 October 2015).

5. http://www.bbc.co.uk/news/uk-13566476 (accessed 30 October 2015).

6. http://natsol.co.uk/ (accessed 30 October 2015).

7. http://www.waterrf.org/PublicReportLibrary/RFR90781_1999_241A.pdf (accessed 30 October 2015).

8. http://www.thegreenage.co.uk/tech/water-saving-showerheads/ (accessed 30 October 2015).

9. http://www.epa.gov/watersense/docs/showerheads_finalsuppstat508.pdf (accessed 30 October 2015).

10. https://www.savewatersavemoney.co.uk/products/view/3056/free-4-minute-shower-timer-united-utilities.html (accessed 2 November 2015).

11. http://www.which.co.uk/energy/creating-an-energy-saving-home/reviews-ns/water-saving-products/water-efficient-washing-machines/ (accessed 30 October 2015).

12. http://www.waterwise.org.uk/pages/water-saving-advice-and-tips.html (accessed 30 October 2015).

13. http://www.greenhotelier.org/know-how-guides/water-management-and-responsibility-in-hotels/ (accessed 30 October 2015).

14. http://www.propelair.com/save/ (accessed 2 November 2015).

15. http://starwoodworldwide.custhelp.com/app/answers/detail/a_id/268/~/make-a-green-choice (accessed 30 October 2015).

16. http://facilityexecutive.com/2012/04/the-hvac-factor-reducing-water-consumption/ (accessed 30 October 2015).

17. http://www.energy.ca.gov/title24/2013standards/prerulemaking/documents/2011-04-27_workshop/review/2013_CASE_CTWS_042011.pdf (accessed 30 October 2015).

18. http://www.mde.state.md.us/programs/Water/WaterConservation/Pages/Programs/WaterPrograms/water_conservation/household_tips/carwashing.aspx (accessed 30 October 2015).

19. http://www.allianceforwaterefficiency.org/Vehicle_Wash_Introduction.aspx (accessed 30 October 2015).

20. http://temp.waterfootprint.org/?page=files/Animal-products (accessed 2 November 2015).
21. http://temp.waterfootprint.org/?page=files/Softdrinks (accessed 2 November 2015).
22. http://www.hsph.harvard.edu/nutritionsource/healthy-drinks/soft-drinks-and-disease/ (accessed 30 October 2015).
23. https://www.generationawake.eu/en/consumption-guide/the-water-guide/ (accessed 30 October 2015).
24. http://www.treehugger.com/sustainable-fashion/10-awesome-innovations-changing-future-fashion.html (accessed 30 October 2015).
25. http://www.data360.org/dsg.aspx?Data_Set_Group_Id=757 (accessed 30 October 2015).

Chapter 11

1. http://www.tn.gov.in/dtp/rainwater.htm (accessed 30 October 2015).
2. http://www.thehindu.com/todays-paper/tp-national/tp-tamilnadu/tamil-nadu-praised-for-its-good-rainwater-harvesting-model/article2495647.ece (accessed 30 October 2015).
3. http://lankarainwater.org/wp/ (accessed 30 October 2015).
4. http://www.oas.org/dsd/publications/unit/oea59e/ch10.htm (accessed 30 October 2015).
5. Galvalume® web site http://www.galvalume.com/ (accessed 30 October 2015).
6. Texas Water Development Board web site https://www.twdb.texas.gov/innovativewater/rainwater/docs.asp (accessed 30 October 2015).
7. http://www.rainharvesting.co.uk/pdfs/case_studies/Adnams.pdf (accessed 30 October 2015).
8. http://www.zdnet.com/article/harvesting-rainwater-for-more-than-greywater/ (accessed 2 November 2015).
9. https://www.health.ny.gov/environmental/water/drinking/coliform_bacteria.htm (accessed 30 October 2015).
10. http://www.hc-sc.gc.ca/ewh-semt/pubs/water-eau/giardia_cryptosporidium-eng.php (accessed 30 October 2015).
11. http://www.hse.gov.uk/legionnaires/ (accessed 30 October 2015).
12. http://www.rainwaterharvesting.co.uk/downloads/farming_environment_agency_guide.pdf (accessed 30 October 2015).
13. http://www.twdb.texas.gov/publications/brochures/conservation/doc/RainwaterHarvestingManual_3rdedition.pdf (accessed 2 November 2015).
14. http://www.breeam.org/ (accessed 30 October 2015).
15. http://www.usgbc.org/LEED (accessed 30 October 2015).
16. http://www.greenkey.global/ (accessed 2 November 2015).
17. http://bristolopeningdoors.org/horizon-house/# (accessed 30 October 2015).
18. http://www.stormsaver.com/Case-Studies/horizon-house-is-the-environment-agencies-national-office-in-bristol-city-centre-/12541 (accessed 2 November 2015).
19. http://www.dreiseitl.com/ (accessed 30 October 2015).
20. http://architecture.mit.edu/class/nature/archive/student_projects/fmr/project_cases_platz.htm (accessed 30 October 2015).

21. http://www.clearwater.asn.au/resource-library/case-studies/royal-park-stormwater-harvesting-project.php (accessed 30 October 2015).

22. http://www.pub.gov.sg/water/Pages/LocalCatchment.aspx (accessed 30 October 2015).

23. http://documents.worldbank.org/curated/en/2006/07/7138489/dealing-water-scarcity-singapore-institutions-strategies-enforcement (accessed 30 October 2015).

24. http://search.informit.com.au/documentSummary;dn=826920873207007;res=IELENG (accessed 2 November 2015).

25. http://www.ksl.com/?sid=4001252 (accessed 30 October 2015).

26. http://www.dcconservation.com/Rainwater%20Harvesting/Rainwater%20Harvesting.htm (accessed 2 November 2015).

27. http://w1.weather.gov/glossary/index.php?word=dew+point (accessed 30 October 2015).

28. http://harialanzarote.com/stone-mulching/ (accessed 30 October 2015).

29. http://goo.gl/Ci44mh (accessed 30 October 2015).

30. http://u.cs.biu.ac.il/~trakht/JAE06.pdf (accessed 30 October 2015).

31. http://blogs.ei.columbia.edu/2011/03/07/the-fog-collectors-harvesting-water-from-thin-air/ (accessed 30 October 2015).

32. http://www.opur.fr/angl/question1_ang.htm (accessed 30 October 2015).

33. http://www.rsc.org/chemistryworld/News/2005/August/31080502.asp (accessed 30 October 2015).

34. http://www.allianceforwaterefficiency.org/Condensate_Water_Introduction.aspx (accessed 30 October 2015).

Chapter 12

1. http://www.pub.gov.sg/about/historyfuture/Pages/NEWater.aspx (accessed 30 October 2015).

2. http://www.fao.org/nr/water/aquastat/countries_regions/qat/index.stm (accessed 30 October 2015).

3. http://news.nationalgeographic.com/news/2007/11/071108-australia-drought.html (accessed 30 October 2015).

4. http://www.thembrsite.com/about-mbrs/what-are-mbrs/ (accessed 30 October 2015).

5. http://voices.nationalgeographic.com/2012/01/16/sewer-mining-coming-to-a-community-near-you/ (accessed 2 November 2015).

6. http://www.motherearthnews.com/green-homes/home-design/greywater-zm0z11zphe.aspx (accessed 2 November 2015).

7. http://www2.epa.gov/nutrientpollution/problem (accessed 30 October 2015).

8. http://info.cat.org.uk/sites/default/files/documents/ts_gw.pdf (accessed 30 October 2015).

9. Judd, S. (2006) *The MBR Book. Principles and Applications of Membrane Bioreactors in Water and Wastewater Treatment*, Elsevier, Oxford.

10. http://appliedmembranes.com/ultrafiltration-membranes-uf-membranes.html (accessed 2 November 2015).

11. http://mtstandard.com/news/opinion/gray-water-law-is-a-good-step-forward/article_fe1337c2-7eab-54b0-bdec-ccf0f05837c1.html (accessed 30 October 2015).

12. http://www.water.nsw.gov.au/urban-water/recycling-water/greywater (accessed 2 November 2015).

13. http://www.thenbs.com/publicationindex/DocumentSummary.aspx?PubID=76&DocID=294669 (accessed 2 November 2015).

14. http://ec.europa.eu/environment/water/water-bathing/index_en.html (accessed 30 October 2015).

15. http://www.breeam.com/page.jsp?id=86 (accessed 2 November 2015).

16. http://web.stanford.edu/group/greendorm/greendorm.html (accessed 2 November 2015).

17. http://www.greywater.com/treatment.htm (accessed 2 November 2015).

18. http://www.afedonline.org/water%20efficiency%20manual/pdf/7appendix%20a_case%20studies.pdf (accessed 2 November 2015).

19. http://www.theguardian.com/lifeandstyle/2014/jul/21/greywater-systems-can-they-really-reduce-your-bills (accessed 2 November 2015).

20. http://www.reaquasystems.com/how-reaqua-works/test-page/ (accessed 30 October 2015).

Chapter 13

1. http://www.un.org/press/en/2010/ga10967.doc.htm (accessed 30 October 2015).

2. http://whqlibdoc.who.int/publications/2011/9789241548151_eng.pdf?ua=1 (accessed 30 October 2015).

3. http://whqlibdoc.who.int/publications/2011/9789241548151_eng.pdf?ua=1 (accessed 30 October 2015).

4. http://www.biosandfilter.org/biosandfilter/index.php/item/320 (accessed 30 October 2015).

5. http://www.who.int/water_sanitation_health/dwq/wsh0207/en/index4.html (accessed 30 October 2015).

6. http://www.who.int/water_sanitation_health/dwq/arsenic/en/ (accessed 30 October 2015).

7. http://water.epa.gov/drink/contaminants/basicinformation/nitrate.cfm (accessed 2 November 2015).

8. http://ec.europa.eu/environment/water/water-nitrates/index_en.html (accessed 30 October 2015).

9. http://www.nbcnews.com/id/23503485/#.VTPKEBPF9Fg (accessed 30 October 2015).

10. http://www.livescience.com/20532-birth-control-water-pollution.html (accessed 30 October 2015).

11. http://www.ukradon.org/ (accessed 30 October 2015).

12. http://www.epa.gov/ogwdw/consumer/pdf/hist.pdf (accessed 30 October 2015).

13. http://water.epa.gov/lawsregs/rulesregs/sdwa/currentregulations.cfm (accessed 30 October 2015).

14. http://ec.europa.eu/environment/water/water-drink/legislation_en.html (accessed 30 October 2015).

15. https://www.chinadialogue.net/article/show/single/en/7722-China-s-drinking-water-safety-faces-scrutiny-in-2-15 (accessed 30 October 2015).

16. http://www.encyclopedia.com/doc/1G2-3407400129.html (accessed 30 October 2015).

17. http://www.ohiohistorycentral.org/w/Cuyahoga_River_Fire?rec=1642 (accessed 30 October 2015).
18. http://ec.europa.eu/environment/water/water-framework/ (accessed 2 November 2015).
19. http://ec.europa.eu/environment/pubs/pdf/factsheets/water-framework-directive.pdf (accessed 30 October 2015).
20. http://www2.geog.ucl.ac.uk/~bpage/files/EuropeanEnvironmentpart2.pdf (accessed 30 October 2015).
21. http://www2.geog.ucl.ac.uk/~bpage/files/EuropeanEnvironmentpart2.pdf (accessed 31 October 2015).
22. Pitt, R., Bannerman, R., Clark, S., Williamson, D. (2005) Sources of pollutants in urban areas (Parts 1 and 2), in *Effective Modeling of Urban Water Systems* (eds Kim N. Irvine, E. A. McBean and Robert E. Pitt). CHI, Guelph.
23. http://naei.defra.gov.uk/overview/pollutants?pollutant_id=17 (accessed 31 October 2015).
24. http://www.independent.co.uk/environment/un-hails-green-triumph-as-leaded-petrol-is-banned-throughout-africa-6112912.html (accessed 2 November 2015).
25. http://www.dec.ny.gov/chemical/67239.html (accessed 31 October 2015).

Chapter 14

1. http://www.theguardian.com/commentisfree/2012/may/08/water-industry-pipes-scandal (accessed 2 November 2015).
2. http://www.planningportal.gov.uk/uploads/code_for_sust_homes.pdf (accessed 31 October 2015).
3. http://www.biosolarroof.com/ (accessed 31 October 2015).
4. http://architizer.com/blog/france-green-roof-law/ (accessed 31 October 2015).
5. http://mitpress.mit.edu/books/fighting-traffic (accessed 31 October 2015).
6. https://workspace.imperial.ac.uk/civilengineering/Public/Technical%20papers%20B/22B-Trioulet-Impact%20of%20Urban%20Creep%20on%20the%20Hydrology%20of%20a%20Catchment.pdf (accessed 2 November 2015).
7. http://www.pps.org/reference/jgehl/ (accessed 31 October 2015).
8. http://www.treesforcities.org/about-us/20th-birthday/ (accessed 31 October 2015).
9. http://www.trees.org.uk/aa/documents/amenitydocs/2013_documents/wed_05_Bjorn_Embren-The_Stockholm_Solution.pdf (accessed 31 October 2015).
10. http://lufa.com/en/ (accessed 31 October 2015).
11. Clay, J. (2004) *World Agriculture and the Environment: A Commodity-by-Commodity Guide to Impacts and Practices*, Island Press, Washington DC.
12. http://waterfootprint.org/media/downloads/Hoekstra-2008-WaterfootprintFood.pdf (accessed 31 October 2015).
13. https://pcrm.org/health/cancer-resources/diet-cancer/facts/meat-consumption-and-cancer-risk (accessed 31 October 2015).
14. http://www.webmd.com/heart-disease/news/20141105/red-meat-carnitine (accessed 31 October 2015).
15. http://urbankayaks.com/collections/kayak-tours (accessed 31 October 2015).
16. http://www.kayakinglondon.com/ (accessed 31 October 2015).

17. http://next.paris.fr/english/visit/flagship-events/paris-plages/rub_8208_stand_34146_port_18969 (accessed 31 October 2015).
18. http://gogermany.about.com/od/berlin/ss/Badeschiff-Berlin.htm (accessed 31 October 2015).
19. http://www.wildlondon.org.uk/reserves/camley-street-natural-park (accessed 31 October 2015).
20. http://www.wwt.org.uk/wetland-centres/london/ (accessed 31 October 2015).
21. http://www.nhm.ac.uk/visit/galleries-and-museum-map/wildlife-garden.html (accessed 31 October 2015).
22. http://www.landscapeinstitute.co.uk/casestudies/casestudy.php?id=1 (accessed 31 October 2015).
23. http://www.spiegel.de/international/spiegel/amphibious-houses-dutch-answer-to-flooding-build-houses-that-swim-a-377050.html (accessed 31 October 2015).
24. http://www.bostonlivingwithwater.org/portfolio/high-street-city-gradually-living-with-water (accessed 31 October 2015).
25. http://bgd.org.uk/ (accessed 31 October 2015).

Index